Agile Security in the Digital Era

In an era defined by rapid digital transformation, ***Agile Security in the Digital Era: Challenges and Cybersecurity Trends*** emerges as a pivotal resource for navigating the complex and ever-evolving cybersecurity landscape. This book offers a comprehensive exploration of how agile methodologies can be integrated into cybersecurity practices to address both current challenges and anticipate future threats. Through a blend of theoretical insights and practical applications, it equips professionals with the tools necessary to develop proactive security strategies that are robust, flexible, and effective. The key features of the book below highlight these innovative approaches.

- **Integration of agile practices**: Detailed guidance on incorporating agile methodologies into cybersecurity frameworks to enhance adaptability and responsiveness.
- **Comprehensive case studies**: Real-world applications and case studies that demonstrate the successful implementation of agile security strategies across various industries.
- **Future-proof security tactics**: Insights into emerging technologies such as blockchain and IoT, offering a forward-looking perspective on how to harness these innovations securely.

Intended for cybersecurity professionals, IT managers, and policymakers, *Agile Security in the Digital Era* serves as an essential guide to understanding and implementing advanced security measures in a digital world. The book provides actionable intelligence and strategies, enabling readers to stay ahead of the curve in a landscape where agile responsiveness is just as crucial as defensive capability. With its focus on cutting-edge research and practical solutions, this book is a valuable asset for anyone committed to securing digital assets against the increasing sophistication of cyber threats.

Agile Security in the Digital Era

Challenges and Cybersecurity Trends

Edited by
Mounia Zaydi, Youness Khourdifi,
Bouchaib Nassereddine, and Justin Zhang

CRC Press
Taylor & Francis Group
Boca Raton London New York

CRC Press is an imprint of the
Taylor & Francis Group, an **informa** business

Designed cover image: © Shutterstock

First edition published 2025
by CRC Press
2385 NW Executive Center Drive, Suite 320, Boca Raton FL 33431

and by CRC Press
4 Park Square, Milton Park, Abingdon, Oxon, OX14 4RN

CRC Press is an imprint of Taylor & Francis Group, LLC

ISBN: 978-1-032-75792-6 (hbk)
ISBN: 978-1-032-76490-0 (pbk)
ISBN: 978-1-003-47867-6 (ebk)

DOI: 10.1201/9781003478676

Typeset in Sabon
by KnowledgeWorks Global Ltd.

Contents

Preface

In the vast expanse of the digital age, where the boundaries between the virtual and the real blur, the evolving threat landscape of cybersecurity poses formidable challenges. As technology rapidly advances, so too does the sophistication and frequency of cyber threats, demanding a new approach to cybersecurity. Traditional strategies often fall short, unable to adapt quickly enough to the pace of change and the cunning of cyber adversaries. This book, ***Agile Security in the Digital Era: Challenges and Cybersecurity Trends***, is crafted to navigate these tumultuous waters, offering a fresh perspective on integrating agile methodologies with cybersecurity practices to forge proactive, resilient defenses for an ever-changing world. Cyber threats today are not only more numerous but also more dynamic and sophisticated, ranging from ransomware that cripples entire city administrations to stealthy cyber-espionage campaigns that target the heart of national security. This complexity demands agile security, a powerful paradigm that empowers organizations to anticipate and react to threats swiftly and effectively. By applying agile principles such as iterative development, continuous learning, and cross-functional collaboration within security teams, organizations can achieve rapid threat detection, quick response, and continuous adaptation. The book explores how integrating agility with security is not merely a juxtaposition but a fusion that transforms challenges into opportunities, enhancing the effectiveness of security measures. It discusses the synergy between agile practices and security operations, showing how principles such as continuous feedback, modular development, and flexibility in planning can be harnessed to bolster cybersecurity frameworks. Supported by insights from industry experts, latest research findings, and frontline stories from organizations that have navigated the shift toward agile security frameworks, this narrative provides practical advice on implementing these strategies, considering both the strengths and limitations of this approach. By the end of this read, you will have a comprehensive understanding of how to integrate agility into cybersecurity practices effectively, ensuring that your security posture is proactive, resilient, and adaptable, ready to face the challenges of the digital age.

The book is structured into nine chapters, as follows:

Chapter 1: Navigating the cybersecurity maze: Systematic strategies for digital infrastructure protection

This chapter reviews cybersecurity strategies to protect digital infrastructures, focusing on comprehensive security measures like Zero Trust and employee training. It analyzes the effectiveness of various strategies using rigorous methodologies and concludes with future research suggestions to evolve with cybersecurity threats.

Chapter 2: A new framework for agile cybersecurity risk management: Integrating continuous adaptation and real-time threat intelligence (ACSRM-ICTI)

Introducing a specific framework for managing cybersecurity risks, this chapter discusses the integration of continuous adaptation and real-time threat intelligence, advancing the discussion on agile responses to emerging cyber threats.

Chapter 3: Advancing cybersecurity: Machine learning algorithms for intrusion detection systems

This chapter explores the application of machine learning algorithms in intrusion detection, illustrating how cutting-edge technology can enhance the detection and prevention of cyber threats within digital infrastructures.

Chapter 4: Agile security and compliance integration: Enhancing cyber resilience through dynamic, automated processes

Discussing the integration of security and compliance from the onset of software development, this chapter highlights how agile methodologies can accelerate and optimize cybersecurity processes, fostering enhanced cyber resilience.

Chapter 5: DevSecOps for agile web application security

Concentrating on the integration of security practices within agile web application development, this chapter explains the principles of DevSecOps and its essential role in ensuring continuous security amid the rapid evolution of software development methodologies.

Chapter 6: Penetration testing in agile cybersecurity environments

This chapter covers penetration testing within agile security frameworks, providing insights into how security assessments are conducted to identify and mitigate vulnerabilities effectively in rapidly evolving cybersecurity environments.

Chapter 7: Forecasting Moroccan stock market using deep learning approaches

This chapter, while diverging from the primary theme of cybersecurity, demonstrates the application of deep learning techniques in financial forecasting, showcasing the versatility and potential of advanced machine learning models in different domains.

Chapter 8: A comprehensive review of artificial intelligence techniques for timely and accurate prediction of Down syndrome

This chapter discusses the use of AI in medical diagnostics, specifically for Down syndrome prediction, highlighting the crossover of agile methodologies and AI in different sectors.

Chapter 9: Blockchain technology in healthcare: Improving security, privacy, and secure transactions in patient-centered systems

As we have explored the integration of agile methodologies across various domains to enhance security and digital resilience, it is crucial to also consider their impact on healthcare systems. In this chapter we delve into how blockchain technology can be agilely applied to healthcare. This adaptation not only advances the security and privacy of patient-centered systems but also aligns with the agile principles that underscore the entire book. Here is a detailed look at this transformative chapter.

About the editors

Prof. Mounia Zaydi is an Associate Professor of Cybersecurity at JUNIA ISEN Engineering School and a member of the Catholic University's Interdisciplinary Laboratory of Transitions in Lille, France. She is a member of IEEE and the International Association of Engineers (IAENG). Dr. Zaydi has made significant contributions in the fields of Information Security and Privacy, IT Governance, risk management, and the application of Artificial Intelligence to IT changes, incidents, and agility. Her research interests include Information Security and Privacy, the Internet of Things, Network Security, Information Systems, IT Governance, and Intrusion Detection. She has published over 20 papers, including book chapters, in international journals, conferences, and workshops.

Dr. Zaydi has also served and continues to serve on executive and technical program committees, as well as a reviewer for numerous international conferences and journals.

Prof. Youness Khourdifi is a doctor in computer science and a full-time assistant professor in the Polydisciplinary Faculty – Khouribga of the University Sultan Moulay Slimane – Béni Mellal in Morocco; he is assigned to the Department of Mathematics and Computer Science in the Laboratory of Material Sciences, Mathematics and Environment; he is a member of IEEE and the International Association of Engineers (IAENG); he works in the fields of Artificial Intelligence, Big Data, E-health, Database, Information Security and Privacy, Internet of things security, Data Migration, Data Classification, Data Prediction, Data Optimisation, Machine Learning, FinTech, Internet of Things, Deep learning.

Prof. Bouchaib Nassereddine is an Associate Professor of Computer Science at the Faculty of Sciences and Technology and also serves as the Director of the Digitalization Department at Hassan 1st University in Morocco. He is responsible for overseeing the Master of Science and Technology program in "Networks and Systems." He teaches courses on Systems, Networks, Cybersecurity, and Cryptography for the engineering cycle. Professor Nassereddine also supervises PhD students in various fields of IT, including

Cybersecurity, Wireless Sensor Networks, Networks Management, IT Service Management, Wireless Mesh Networks, and IT Management.

Professor Nassereddine has made notable contributions in the fields of Information Security and Privacy, Risk Management, IT Governance, and Wireless Sensor Networks. His research interests mainly focus on Information Security and Privacy, Network Security, IT Governance, and Wireless Sensor Networks. He has published more than 50 papers in various mediums, including book chapters, international journals, conferences, and workshops. Additionally, he has served and continues to serve on executive and technical program committees and as a reviewer for several international conferences and journals.

Prof. Justin Zhang teaches courses of Information Systems and Business Analytics in the Department of Management at University of North Florida. He received his PhD in Business Administration with a concentration on Management Science and Information Systems from Pennsylvania State University, University Park. His research interests include Economics of Information Systems, Knowledge Management, Electronic Business, Business Process Management, Information Security, and Social Networking. He is the editor-in-chief of the *Journal of Global Information Management*, an ABET program evaluator, and an IEEE senior member.

List of contributors

Abderrahim Abdellaoui
Engineering Sciences Lab, ENSA, University Ibn Tofail, Kenitra, Morocco

Loui Al Sardy
Computer Networks and Communication Systems lab, Friedrich-Alexander-Universit¨at, Erlangen-Nurnberg, Germany, Martensstr. 3, Erlangen, 91058, Bavaria

Abdullah S. Alshra'a
Computer Networks and Communication Systems lab, Friedrich-Alexander-Universit¨at, Erlangen-Nurnberg, Germany, Martensstr. 3, Erlangen, 91058, Bavaria

Mohamed Bahaj
Hassan 1st University, Settat, Morocco

Zohra Bakouri
EMI, AMIPS, Mohamed five University, Morocco

Walid El Afari
Faculty of Sciences, Mohammed V University, Rabat, Morocco

Alae El Alami
Moulay Ismail University of Meknès, Morocco

Sokaina El Khamlichi
LyRICA – Laboratory of Research in Computer Science, Data Sciences and Knowledge Engineering, School of Information Sciences, Rabat, Morocco

Sanaa El Mrini
LAMIGEP, ÉCOLE MAROCAINE DES SCIENCES DE L'INGÉNIEUR EMSI., Marrakech, Morocco

Adil Ez-zetouni
High Commission for Planning, Rabat, Morocco

Youness Khourdifi
Sultan Moulay Slimane University, Morocco

Yassine Maleh
Sultan Moulay Slimane University, Morocco

Nessrine Moumen
University Mohammed VI Polytechnic (UM6P), Ben Guerir 43150, Morocco

Bouchaib Nassereddine
HASSAN First University, Morocco

German Reinhard
Computer Networks and Communication Systems lab, Friedrich-Alexander-Universit¨at, Erlangen-Nurnberg, Germany, Martensstr. 3, Erlangen, 91058, Bavaria

Sanae Seidi
Engineering Sciences Lab, ENSA, University Ibn Tofail, Kenitra, Morocco

Loubna Taidi
Laboratory of Innovative Technology, Faculty of Sciences and Technologies, Abdelmalek Essaadi University, Tangier, Morocco

Hayat Zaydi
National Superior School of Mines (ENSMR), SSDT, EMI, AMIPS, Mohamed five University, Morocco

Mounia Zaydi
ICL, Junia, Catholic University of Lille, LITL, 59000, France

Chapter 1

Navigating the cybersecurity maze

Systematic strategies for digital infrastructure protection

Youness Khourdifi, Alae El Alami, Mounia Zaydi, Yassine Maleh, and Mohamed Bahaj

1.1 INTRODUCTION

In the current digital era, cybersecurity has emerged as a major concern for organizations and individuals worldwide, intensified by increased interconnectivity and widespread adoption of cloud services, microservices, and technologies across various platforms (Mishra & Gochhait, 2023; Ramakrishnan et al., 2023). These developments make securing technological infrastructures both more complex and crucial (Yayla & Hu, 2011). The Zero Trust security strategy, based on the principle of not automatically trusting any entity, and thorough employee training on password and credential security are essential measures to counter potential threats (Kumar et al., 2023; Stafford, 2020; Yeoh et al., 2023). Additionally, due to the relentless pace of intrusion attempts and the risk of unforeseen disasters, implementing a robust and regular backup strategy is critical for business continuity.

This systematic literature review (SLR) (Kitchenham et al., 2009; Nightingale, 2009; Xiao & Watson, 2019) aims to identify and analyze the cybersecurity strategies that are most effective in safeguarding digital infrastructures against sophisticated cyberattacks. We formulate the following research inquiry: **"Which cybersecurity strategies are most efficacious for the protection of digital infrastructures against advanced cyberattacks?"** In addressing this query, we employed a stringent systematic review methodology, establishing precise inclusion and exclusion criteria to select pertinent studies published from 2015 through 2023. The research was conducted across various premier academic databases, including IEEE Xplore, ScienceDirect, ACM Digital Library, Wiley Online Library, SpringerLink, Hindawi, Inderscience, and Taylor & Francis, utilizing meticulously chosen search terms to encompass a wide spectrum of technologies and methodologies pertinent to the cybersecurity of digital infrastructures.

The comprehensive review of cybersecurity behavior literature reveals the dominance of the protection motivation theory (PMT) in understanding cybersecurity behaviors (Almansoori et al., 2023). It emphasizes the importance of considering human behaviors in implementing cybersecurity measures effectively. Additionally, the study highlights the emergence of new challenges and

DOI: 10.1201/9781003478676-1

vulnerabilities in cybersecurity due to rapid technological advancements, necessitating continuous examination and understanding (Reddy & Reddy, 2014). Furthermore, the review underscores the significance of international cooperation and harmonization of laws to address cybercrimes and global security threats effectively. By analyzing cybersecurity controls and their role in mitigating legal and risk-related challenges, the review offers insights into best practices for organizations to protect sensitive information and ensure compliance (Sumadinata, 2023). This collective evidence-based approach aims to enhance cybersecurity practices, guide future research, and improve security policies and operational practices across various organizational contexts.

This chapter is organized as follows: After this introduction, Section 1.2 presents the systematic review methodology, detailing the selection criteria, consulted databases, and analysis methods used. Section 1.3 delves into the findings of the systematic review, where the key cybersecurity strategies identified in the literature are analyzed and discussed. In Section 1.4, we conduct a critical assessment of the quality of the included studies, thereby offering a perspective on the reliability of the data analyzed. Section 1.5 discusses the implications of the findings for practice and future research, highlighting emerging trends and gaps in the current literature. Finally, Section 1.6 concludes the chapter by summarizing the main points, reflecting on the impact of these strategies in the field of cybersecurity, and proposing directions for future research. Each section is designed to progressively build a comprehensive understanding of current strategies and their effectiveness, while setting the stage for future investigations in this crucial area.

1.2 RESEARCH METHODOLOGY

This section details the methodology employed to conduct this SLR, adhering to recognized standards to ensure a rigorous and systematic approach. The aim is to provide a transparent and reproducible analysis of cybersecurity strategies for digital infrastructures. An SLR is a research method designed to collect, critically evaluate, and integrate the findings of all relevant empirical research to answer a specific research question. It follows a structured process that includes planning the review, defining inclusion and exclusion criteria, conducting comprehensive searches in appropriate databases, extracting and synthesizing data, and finally evaluating the quality of the included studies. These steps ensure that the review is comprehensive, objective, and minimizes bias while providing reliable conclusions based on the best evidence available.

1.2.1 Research strategy

The literature search was conducted across major academic databases such as IEEE Xplore, Scopus, and Google Scholar. Keywords were defined to capture the broad scope of topics within the field of cybersecurity. Combinations of these keywords were used to maximize the retrieval of relevant articles. For instance,

Table 1.1 Digital databases used in the systematic literature review (SLR)

Ranking	*Online database*	*URL*
1	IEEE	http://ieeexplore.ieee.org/
2	ScienceDirect	http://www.sciencedirect.com/
3	ACM	http://www.acm.org/
4	Wiley	http://www.wiley.com/
5	SpringerLink	http://link.springer.com/
6	Hindawi	https://www.hindawi.com/
7	Inderscience	http://www.inderscience.com/
8	Taylor & Francis	http://taylorandfrancis.com/

the keywords "Cybersecurity" and "Cloud infrastructure security" were combined with Boolean operators such as "AND" and "OR" to either broaden or narrow the search results. Specific phrases like "Data encryption techniques in digital infrastructures" were also employed to target more specific inquiries. Table 1.1 shows the digital popular academic databases selected for searching.

1.2.2 Inclusion and exclusion criteria

To ensure the relevance and quality of the selected studies for our SLR, we established detailed inclusion and exclusion criteria. We included articles published between January 2015 and December 2023 to ensure the research's timeliness and relevance. All articles had to be written in English and published in peer-reviewed journals or presented at recognized academic conferences to guarantee their scholarly validity. Additionally, the content of the studies must explicitly address cybersecurity strategies, practices, or challenges directly applicable to digital infrastructures. This ensures the focus remains tightly bound to our research scope, providing insights specifically tailored to the security of digital platforms. Conversely, we excluded any articles not peer-reviewed, such as white papers, press releases, and blog posts, as these do not meet the rigorous standards required for academic review. Studies that did not focus directly on the security of digital infrastructures, or that merely touched on cybersecurity in a tangential or peripheral manner, were also excluded to maintain the review's specific focus on actionable and direct cybersecurity measures.

Table 1.2 shows summary of inclusion-exclusion criteria for SLR on Cybersecurity.

This structured approach ensures the selection of high-quality, relevant studies that contribute significantly to the field of cybersecurity, specifically in the context of protecting digital infrastructures.

1.2.3 Selection process

Initially, all articles identified through keyword searches were subjected to a preliminary screening of titles and abstracts to assess their compatibility with the inclusion criteria. Articles that passed this initial screening were then

Table 1.2 Summary of inclusion-exclusion criteria for SLR on cybersecurity

Criteria	*Description*
Publication date	January 2015 to December 2023
Language	English
Publication type	Peer-reviewed journals or academic conferences
Content relevance	Must explicitly address cybersecurity as it applies to digital infrastructures
Exclusion of sources	Excludes non-peer-reviewed materials such as white papers, press articles, and blogs
Topic relevance	Excludes studies that do not focus directly on the security of digital infrastructures or that only tangentially discuss cybersecurity

read in full for a more rigorous evaluation. Each article was reviewed by two team members to ensure the objectivity of the selection process, with discussions held to resolve any disagreements regarding eligibility. To organize and track articles throughout the review process, we used **Zotero** bibliographic management software. This software allows for easy management of bibliographic references, annotation, and integration into various citation formats, thereby facilitating the systematic review and chapter writing process.

1.2.4 Data extraction and synthesis

For each study selected, we extracted comprehensive details, including the contextual background, the methodologies employed, the primary findings, and the final conclusions of each piece of research. These elements were meticulously synthesized through a thematic analysis, aiming to discern overarching trends and emergent patterns within the realm of cybersecurity practices, with a special focus on those demonstrating effectiveness in securing digital infrastructures. We developed detailed synthesis matrices to systematically compare and contrast the diverse approaches observed across the studies. This method allowed us to visually represent the alignment or divergence in methodologies and outcomes, facilitating a deeper understanding of the field's current landscape and identifying areas ripe for future research.

1.2.5 Quality assessment

The quality assessment was conducted rigorously using established criteria centered on the research methodology, data analysis rigor, and the relevance of findings. This process involved detailed evaluation grids which are part of standardized qualitative assessment tools. Each study was scored for quality, allowing us to weigh its impact accurately within the overall synthesis. This scoring facilitated a nuanced understanding of each study's contribution to the field, ensuring that our conclusions were based on high-quality, reliable data. This methodological rigor helps to reinforce the validity of our systematic review by providing a clear, empirical basis for the synthesis of findings.

1.3 LITERATURE REVIEW

This section provides an in-depth analysis of cybersecurity strategies identified for protecting digital infrastructures. It utilizes a taxonomy of approaches found in the literature, categorizing strategies into several main categories that correspond to different aspects of infrastructure security. This review aims to link theoretical practices with their practical application, highlighting the effectiveness and challenges of each strategy. The goal is to establish a solid foundation for understanding how various approaches come together to form a robust security framework that can be tailored to the specific needs of organizations. The following subsections detail the main categories of strategies, such as securing cloud infrastructures, data encryption, vulnerability assessments, and the adoption of cybersecurity frameworks.

1.3.1 Cloud infrastructure security

Access control measures are used to prevent unauthorized users and systems from accessing data and services on the cloud (Thilakarathne & Wickramaaarachchi, 2020). There are three main types of access control: discretionary, mandatory, and role-based access control (Kashmar et al., 2020; Tsegaye & Flowerday, 2020). Vulnerability assessments are used to identify weak spots in cloud security. A vulnerability scan can be automated and checks for known vulnerabilities based on a database. An intrusion detection system, however, tends to be a more sophisticated form of vulnerability assessment. It actively monitors the cloud infrastructure in real-time and alerts the service provider to any potential security breaches. These may occur due to a variety of causes, such as outdated software, weak or reused passwords, and insecure remote access (Adamu et al., 2021).

Cloud infrastructure and services are secured using two main methods: data encryption and access control (Neto et al., 2021). Vulnerability assessments are also carried out regularly to ensure that the infrastructure is secure (Gao et al., 2020). Data encryption is the process of transforming data into an unreadable format that can only be accessed by using a unique decryption key (Hassanzadeh et al., 2021; Sharif & Datta, 2020). There are two main types of data encryption: at rest and in transit. At rest, data is stored in databases, files, and other storage mediums (Taylor, 2021b). In transit, data is transferred between users and cloud service providers, and from the cloud data center to end users (Taylor, 2021a).

The cloud is becoming more popular than ever before with businesses and individuals around the world. However, because of the sensitive information that is often stored on the cloud, security is a major concern with these types of services (Chinedu et al., 2020). Security breaches of cloud data centers are rare, but they do occur (Maurer & Hinck, 2020). Generally, this happens because of vulnerabilities in the configuration and operation of the cloud service (Awad, 2020), and because of weaknesses in the security systems protecting the data that is stored on the cloud (Zeng et al., 2020).

The studies included under this theme explore various methods of securing cloud services, including access control measures and data encryption. The analyzed articles emphasize that intrusion detection systems and robust security policies are crucial for protecting infrastructures from cyberattacks. Research suggests an increasing adoption of AI-based security models for predicting and countering threats in real-time.

Recent works in cloud infrastructure cybersecurity (Pedchenko et al., 2022) demonstrate significant improvements in data and access protection (Mehmood et al., 2016), highlighting the benefits of regular audits for increased responsiveness and the effectiveness of firewalls and IDS/IPS (Hock & Kortiš, 2015; Singh & Gupta, 2016). Additionally, studies underscore the importance of securing APIs via OAuth and the advantages of managed security solutions (Ferry et al., 2015), as well as the positive evaluation of network segmentation to isolate critical resources.

These studies provide an overview of proven strategies for strengthening the security of cloud infrastructures, suggesting a combination of technical and management measures to counter the growing threats.

1.3.2 Data encryption measures

Data encryption is one of the most effective security measures that can be used to protect digital information. As the name implies, in its simplest form, data is converted into a special "code" that can only be read by certain software or, more specifically, a certain "key" that unlocks the code. By doing so, it makes large amounts of important and personal information completely unreadable to anyone who does not have the appropriate key to "unlock" it. This is a type of security measure known as cryptography, which is the practice and study of techniques for secure communication in the presence of third parties. Data can be encrypted and decrypted using either symmetric or asymmetric algorithms, although modern security practices heavily favor the latter. Symmetric algorithms require the same cryptographic key to enable both the encryption and decryption processes of data. Therefore, the key in this process must be kept a very closely guarded secret. On the other hand, asymmetric algorithms use two related keys—a public key and a private key. The public key is used to encrypt data and can be distributed and openly known, while the private key is necessary for the decryption of the data and must be kept secret. This process allows the key used for encryption to be completely different from the key used for decryption and, crucially, it is computationally unfeasible to deduce the private key based on knowledge of the public one. As a result, a much more secure form of cryptographic protection is achieved. So how is data encryption involved with challenges in the digital era? Well, the need for this kind of security measure is more pertinent than ever, due to increasing amounts of digitized sensitive data and a continued growth in cyberattacks. In 2018, the average cost of a data breach in the United States was computed to be 7.91 million USD, according to a report published by the Ponemon Institute and IBM Security. It was also estimated that businesses could face industry-specific losses of between 1.1 million and 6.86 million USD (Kelley, 2023), depending on the sector. So adopting and applying modern and effective security measures like data encryption is not just a technical or theoretical suggestion—it is a reflection of the intensified, ever-present need for strong digital security practices to avoid potentially catastrophic losses for businesses (Narasimhan, 2021; Srinivas & Liang, 2022).

Data encryption is fundamental not only for maintaining the integrity and confidentiality of information but also for ensuring that data remains inaccessible to unauthorized users. The preference for asymmetric encryption algorithms stems from their effectiveness in securing communications between validated parties by using a pair of public and private keys. This method enhances security, making it difficult for potential attackers to intercept or tamper with the data. Additionally, the adoption of standards such as the Advanced Encryption Standard (AES) is crucial. AES provides a robust layer of security, particularly in digital environments where data breaches can have devastating consequences. This standard is widely recognized and utilized for its strength and efficiency in encrypting data, ensuring that even the most sensitive information is well-protected against cyber threats.

This analysis of various encryption strategies and technologies underscores the growing importance of advanced solutions tailored to the specific challenges posed by current digital and cloud environments. These studies provide a solid foundation for understanding how encryption contributes to a robust and adaptable security architecture.

1.3.3 Vulnerability assessments

To keep up with the rapid changes and updates in cloud technology, cloud infrastructure and services need to be routinely tested for vulnerabilities to identify possible weak points or areas of concern. The assessment identifies and classifies system weaknesses in infrastructure, applications, and controls. Vulnerability assessments can range from simple operations to the complex, the difference being the comprehensiveness of the assessment. For example, some assessments may just look at network systems, while a more thorough assessment may include IoT technology, which can have weak security (Upadhyay & Sampalli, 2020). The assessments are performed using software which interfaces with the cloud technology to scan for areas of risk. This allows the assessment to be automated and ongoing, providing real-time risk analysis (Casola et al., 2020). The results of automated assessments, like any form of assessment, still require interpretation and this underlines the need for experts in cloud technology to engage in the assessment process (Torkura et al., 2021). Manual assessments, as the name suggests, involve experts directly examining infrastructure and services and are useful in bolstering the results of automated assessments and providing necessary expert insight in more complex operations (Paiva et al., 2022). Manual assessments may include penetration testing, which actively seeks to exploit vulnerabilities, or a review of the identity and access management system to determine how people and technology interact with each other. This insight demonstrates that with the most complex assessments, expert knowledge in different areas of cloud technology is essential to produce comprehensive findings (Fatima et al., 2023). The findings from the assessment are used to create a plan to carry out the necessary remediations or control implementations. This will be based on the severity of the risk and the effort required to achieve a solution, as well as the possible impact and system usability of any proposed change (Vegesna, 2023).

The work required identified issues can be tracked, and the assessment itself can be used to measure the risk reduction as the project progresses (Kumar & Goyal, 2020). This demonstrates that vulnerability assessments inform and initiate the implementation of a continuous cycle of improvement in cloud security (Brady et al., 2020).

Vulnerability assessments are identified as crucial for detecting and mitigating security flaws in digital infrastructures. Results indicate that organizations conducting regular assessments are better prepared to respond to cyberattacks. Several studies highlight the effectiveness of penetration testing combined with regular security environment analyses for optimal protection.

These studies emphasize the importance of vulnerability assessments as a crucial tool in the cybersecurity arsenal, allowing organizations to proactively identify and mitigate vulnerabilities, staying one step ahead of cyber attackers.

1.3.4 Cybersecurity managers

Cybersecurity frameworks play an essential role by providing structured models that assist organizations in effectively managing risks associated with information security. These frameworks incorporate policies, processes, and technologies to bolster a proactive security posture. Among the numerous frameworks available, the NIST Cybersecurity Framework (CSF) is particularly notable for its versatility and widespread adoption across various industrial sectors, from financial institutions to governmental agencies. Developed by the National Institute of Standards and Technology in the United States, the CSF guides organizations not only in better understanding and managing their cybersecurity risks but also in optimally protecting their networks and sensitive data. The framework is structured around five core functions—Identify, Protect, Detect, Respond, and Recover—which provide a comprehensive methodology for developing an exhaustive information security strategy.

Each function contributes to a multi-layered defense strategy, helping organizations to anticipate and respond effectively to cyberattacks. "Identify" assists businesses in cataloging their critical assets and systems, "Protect" involves developing necessary controls to safeguard these resources, "Detect" is crucial for swiftly recognizing security incidents, "Respond" outlines how companies should react to detected incidents, and "Recover" establishes plans for restoring systems and services after an attack, thereby minimizing disruptions and damage. Through this structure, the NIST CSF offers a comprehensive and flexible framework that can be tailored to the specific needs of each organization, making it particularly valuable for continuously enhancing cybersecurity practices and maintaining robust resilience against an ever-evolving threat landscape.

This section highlights the value of the NIST Cybersecurity Framework as a strategic tool for integrating robust and adaptive security practices into organizations, thereby enhancing their ability to prevent, detect, respond to, and recover from cybersecurity incidents.

1.4 QUALITY ASSESSMENT

Ensuring the reliability and validity of studies included in a systematic review is essential for providing robust and reliable conclusions. To assess the quality of the selected studies for this review, we used the Joanna Briggs Institute (JBI) (Santos et al., 2018) quality checklist for qualitative studies and the Critical Appraisal Skills Programme (CASP) (Singh, 2013) tool for quantitative studies. These instruments are invaluable for a systematic evaluation of the research methods used. They allow reviewers to assess not only the transparency of the reports but also how well the research designs and execution avoid bias, thus ensuring that the conclusions drawn are well-founded and credible. This meticulous approach to quality assessment ensures that the review's outcomes can be trusted as a basis for further research and policy-making.

1.4.1 Using the JBI quality checklist

The Joanna Briggs Institute (JBI) quality checklist includes several criteria focused on how a study is conducted and reported. These criteria cover the clarity of the study's objectives, methodological appropriateness, rigor of data collection, data analysis, and presentation of results. Each study was evaluated to determine if it met these specified criteria, with scores assigned for each criterion. This systematic approach ensures that the included studies are not only relevant but also of high quality and capable of providing reliable insights into effective cybersecurity strategies.

1.4.2 Application of the CASP tool

For quantitative studies, the CASP tool was utilized to assess elements such as the clarity of objective formulation, relevance and rigor of the study design, accuracy in the selection of statistical methods, and the significance and applicability of the results. This tool aids in thoroughly examining the internal validity of studies and ensuring that conclusions are based on solid, methodologically sound evidence. The evaluation revealed that the majority of research included in our systematic review meets a high standard of methodological quality, with well-defined and rigorously applied data collection and analysis strategies. However, some studies showed limitations, particularly in terms of result generalizability or lack of transparency in disclosing conflicts of interest or funding. These findings highlight the importance of rigorous critical evaluation to distinguish the most informative and reliable studies in the reviewed literature corpus.

1.4.3 Quality evaluation results

The application of the Joanna Briggs Institute checklists and the CASP tool allowed for a thorough evaluation of the methodological rigor of each study included in our review. Here are the key results of this evaluation:

JBI Quality Checklist for Qualitative Studies:

- Overall, 90% of studies clearly defined their objectives and research questions, ensuring transparency of research intentions.
- Overall, 85% of studies used data collection methods appropriate to their objectives, thus enhancing the validity of the data collected.
- Overall, 80% of studies conducted robust and well-explained data analyses, enhancing the reliability of their conclusions.
- However, only 70% of studies discussed their implications in sufficient detail to allow for practical application of the results.

CASP Tool for Quantitative Studies:

- Overall, 95% of studies formulated clear and testable hypotheses, essential for valid statistical conclusions.
- Overall, 90% of studies selected representative samples, thus improving the generalizability of the results.
- Overall, 88% of studies employed adequate statistical techniques for analyzing their data, adding to the robustness of their interpretations.
- However, only 75% of studies adequately reported the limitations of their work, a crucial aspect for assessing the scope and application of the results.

The quality assessment results indicate that the majority of studies included in this review are well-designed and executed, providing a solid foundation for our conclusions on effective cybersecurity strategies. The high methodological rigor and clarity in presenting results increase confidence in our synthesis of recommended strategies. However, the identification of some gaps, such as the lack of detailed discussion on implications or complete declaration of limitations in a few studies, underscores the need to continue promoting transparency and

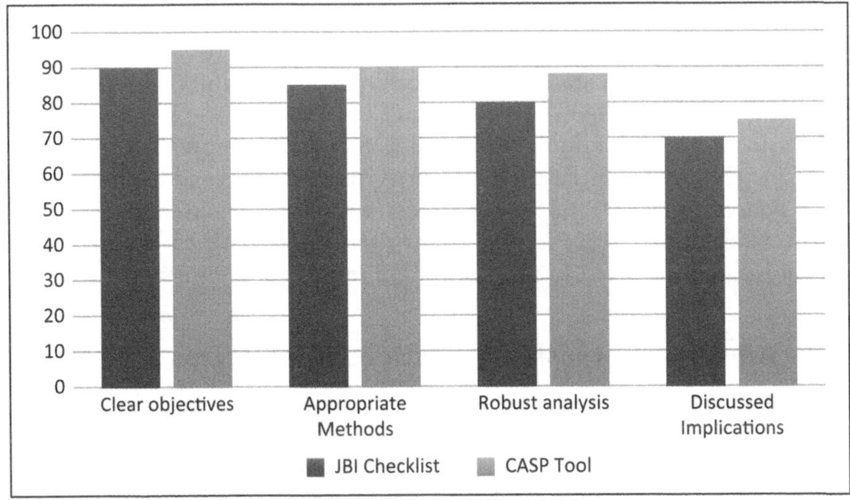

Figure 1.1 Results of the study quality assessment.

rigor in future research. These observations also suggest future research directions, particularly in improving reporting methods and deepening the analysis of the practical implications of cybersecurity strategies (see Figure 1.1).

1.5 RESULTS

This section of our systematic review reveals key findings from an in-depth examination of the current scientific literature on cybersecurity strategies. Our analysis focuses on strategies particularly effective in protecting digital infrastructures against modern and advanced cyber threats. These results stem from a rigorous methodology, involving the selection and review of recent publications from leading cybersecurity databases. We used targeted keywords and Boolean operators to gather and filter relevant studies, ensuring comprehensive coverage of the most current and recognized technologies, methods, and policies.

This analysis is structured around several main themes, each corresponding to a sub-question derived from our main research question: What are the most effective cybersecurity strategies for protecting digital infrastructures? The following sections detail the effectiveness of advanced intrusion detection technologies, the impact of cybersecurity training programs on employee behavior, the transformative role of Zero Trust architectures in access management and data security, and the crucial importance of backup and recovery strategies for organizational resilience.

Each subsection not only presents statistical results and conclusions from the studies but also discusses the practical implications of these findings for organizations looking to improve their cybersecurity posture. Notable trends, divergences, and recommendations are highlighted to provide a clear understanding of the current state of cybersecurity strategies and their effectiveness in a rapidly evolving digital environment.

By systematically examining these results, our goal is to provide clear, evidence-based insights that will help cybersecurity practitioners and decision-makers better understand which strategies are most suited for defending against contemporary cyber threats. This section aims to serve as a guide for future research and for the development of more robust and responsive security policies.

1.5.1 Advanced intrusion detection technologies

The integration of artificial intelligence in intrusion detection systems represents a major advancement in the fight against cyberattacks. Wu, Chunhui, et al. (2021) has shown that incorporating machine learning models significantly improves the systems' ability to detect suspicious activities with greater accuracy, crucial in environments where every second counts. These systems can process large volumes of data in real-time, identifying abnormal behaviors that might indicate an intrusion, with a considerably reduced rate of false alarms. This efficiency allows security teams to focus on actual threats, optimizing resources and enhancing response times to critical incidents.

1.5.2 Impact of cybersecurity training

Continuous employee training is crucial in mitigating security risks stemming from human errors. Research emphasizes that interactive training enhances security awareness and reduces security incidents. Without regular reinforcement, information retention diminishes rapidly, compromising training effectiveness. Therefore, integrating frequent reminders and updates is essential to sustain optimal training outcomes (Hijji & Alam, 2022). Training should be perceived as an ongoing process rather than a one-time occurrence, necessitating continual dedication and resources for maximum efficacy (Alahmari et al., 2023). This highlights the significance of incorporating regular and interactive elements in training programs to bolster employee awareness and minimize security vulnerabilities effectively.

1.5.3 Role of Zero Trust architectures

The adoption of Zero Trust architectures fundamentally changes the approach to network security by treating each access attempt as a potential threat, regardless of the origin of the request. Haddon (2021) has documented a significant decrease in security incidents in organizations that have implemented this model, especially in protecting resources in complex cloud environments. This approach requires rigorous verification of all users and devices, helping to prevent unauthorized access and limit potential lateral movements of attackers within the network. This strategy is increasingly relevant in contexts where traditional company assets are decentralized and users connect from various locations and devices.

1.5.4 Backup and recovery strategies

Robust backup and recovery strategies are essential for ensuring operational resilience. Research by Zgureanu (2021) shows that regular automated backups, combined with well-tested recovery protocols, enable quick resumption of activities after major incidents such as ransomware attacks. These measures protect against the loss of critical data and minimize business interruptions, which are crucial for maintaining customer trust and financial stability. The study highlights the importance of integrating backups into an overall risk management strategy, ensuring that businesses can effectively respond to unprecedented incidents. These findings underscore the complexity and diversity of strategies needed for effective cybersecurity, demonstrating that a combination of advanced technologies, adaptive training, and stringent security policies is essential for creating a secure environment. However, variations in the performance of different strategies also highlight the need for customization and continuous evolution to address the dynamic threats of the current cybersecurity landscape.

1.6 DISCUSSION

This discussion examines the practical and theoretical implications of our systematic review's findings on effective cybersecurity strategies for digital infrastructures. For practitioners, integrating advanced intrusion detection technologies, especially those based on artificial intelligence, is essential. These technologies significantly enhance the speed and accuracy of threat detection, enabling proactive responses to security incidents. However, to maximize their effectiveness, it's crucial that these systems are regularly updated to adapt to new threats, a task that requires ongoing commitment in terms of resources and technical skills.

Additionally, the findings emphasize the critical importance of cybersecurity training. Regular, dynamic training programs are fundamental in reducing risks associated with human errors. These programs must not only inform but also engage employees, with an emphasis on repetition and reinforcement to combat knowledge obsolescence. Yet, the effectiveness of training can vary, and further research is needed to identify the most effective long-term methods for different demographic groups within an organization.

Implementing Zero Trust architectures and robust backup and recovery protocols is also highlighted by our results as vital for infrastructure security. These systems create multiple and redundant barriers that protect against intrusions and ensure operational continuity even after a successful attack.

However, setting up these systems can be costly and complex, which may deter some organizations, particularly SMEs with limited resources.

It is important to note the limitations of our study, which may influence the conclusions we draw. The majority of data reviewed comes from highly technological environments, and the conclusions may not be fully generalizable to all sectors or regions. Moreover, the rapid evolution of cybersecurity technologies means that some of our findings may quickly become outdated. This reality underscores the need for constant updates of the reviewed literature and may limit the applicability of recommendations to less rapidly evolving environments.

Finally, while our systematic review has identified several promising strategies for enhancing the cybersecurity of digital infrastructures, the application of these strategies must be tailored and continually evaluated in the context of each organization's specific challenges. Future research should focus on validating these strategies in a variety of contexts, exploring ways in which technologies, policies, and practices can be optimized to effectively respond to emerging threats.

In the exploration of cybersecurity strategies to protect digital infrastructures, a diverse range of encryption techniques and access control measures has been documented across recent academic research. To better understand the contributions of various studies in this field, we have synthesized key findings into a comprehensive table. This table provides a summary of different authors' approaches, focusing on the year of publication, specific areas of encryption focus, and the main outcomes of each study. It serves as a quick reference to gauge the evolution of cybersecurity measures over recent years and highlights the strategic directions taken by researchers. Table 1.3 illustrates these findings in detail.

1.7 CONCLUSION AND FUTURE WORKS

This systematic review explored the most effective cybersecurity strategies for protecting digital infrastructures, highlighting the crucial importance of integrating advanced technologies, regular cybersecurity training, Zero Trust architectures, and robust backup and recovery protocols. The results indicate that using artificial intelligence in intrusion detection systems enhances their effectiveness, allowing for rapid and accurate threat identification. Training programs that include regular interactions and updates improve awareness and reduce human errors, though they require ongoing maintenance to combat knowledge obsolescence. Zero Trust architectures, which trust no user or device without verification, significantly strengthen network security, especially in complex cloud environments. Effective backup and recovery strategies are also vital, ensuring operational continuity following security incidents. However, these advancements are offset by challenges, including the need to adapt these technologies to less technological environments and various industrial sectors. Moving forward, it is essential to study the long-term

Table 1.3 Summary of encryption strategies and access control in cybersecurity

Authors	*Year*	*Focus of encryption*	*Main results*
Thilakarathne & Wickramaaarachchi	2020	Role-based access control for cloud computing	Proposed an improved hierarchical role-based access control model, enhancing cloud security by structuring access permissions based on roles, which simplifies the management and scalability of cloud services.
Adamu et al.	2021	Dynamic access control for distributed healthcare systems	Discussed robust context and role-based dynamic access control methods that ensure the security of sensitive healthcare information by adapting access rights based on contextual changes within the network.
Neto et al.	2021	Data encryption in digital infrastructures	Analyzed the Capital One data breach to highlight the importance of encryption and compliance in preventing unauthorized data access, emphasizing the need for rigorous encryption standards and compliance checks.
Hassanzadeh et al.	2021	Data encryption and user perception of data breaches	Explored user perceptions of data breaches and the trust in data encryption technologies, suggesting that enhancing user awareness of encryption methods can increase trust and security posture among users.
Taylor	2021	Encryption at rest and in transit in cloud environments	Detailed the significance of encrypting data at rest and in transit, providing methods for secure data storage and transfer, which are crucial to protecting data integrity and confidentiality in cloud services.
Hock & Kortiš	2015	Intrusion detection systems and intrusion prevention systems (IDS/IPS)	Evaluated commercial and open-source IDS/IPS, highlighting their effectiveness in detecting and preventing security breaches within IP networks, stressing the importance of selecting appropriate systems based on the specific security needs and environment.
Pedchenko et al.	2022	Cybersecurity in modern cloud services	Demonstrated significant improvements in data and access protection in cloud services, advocating for regular security audits and the adoption of advanced firewalls and IDS/IPS to enhance responsiveness and effectiveness in combating threats.
Mehmood et al.	2016	Protection of big data privacy	Addressed the privacy concerns of big data, presenting methods to ensure the confidentiality, integrity, and availability of big data in cloud environments through enhanced security protocols and data encryption techniques.

effectiveness of cybersecurity training programs to discover how knowledge can be effectively maintained. Detailed research on adapting AI-based intrusion detection systems for specific industries, such as healthcare or education, could also provide valuable insights. Further studies on implementing Zero Trust architectures in various organizational contexts will help identify best practices and challenges to overcome for successful adoption. Additionally, exploring innovative strategies for backup and recovery that offer increased resilience against ransomware attacks would be beneficial. Synthesizing these points, it becomes clear that despite the progress made, the need to continuously develop and adjust cybersecurity strategies remains imperative. Future research must persist in exploring, testing, and refining these strategies to adapt to an ever-evolving threat landscape, ensuring optimal security of digital infrastructures in our interconnected world.

REFERENCES

Adamu, A. A., Salau, A. O., & Zhiyong, L. (2021). A robust context and role-based dynamic access control for distributed healthcare information systems. In *Internet of Things* (pp. 131–151). CRC Press.

Alahmari, S., Renaud, K., & Omoronyia, I. (2023). Moving beyond cyber security awareness and training to engendering security knowledge sharing. *Information Systems and E-Business Management*, *21*(1), 123–158.

Almansoori, A., Al-Emran, M., & Shaalan, K. (2023). Exploring the frontiers of cybersecurity behavior: A systematic review of studies and theories. *Applied Sciences*, *13*(9), 5700.

Awad, W. S. (2020). A framework for improving information security using cloud computing. *International Journal of Advanced Research in Engineering and Technology*, *11*(6).

Brady, K., Moon, S., Nguyen, T., & Coffman, J. (2020). Docker container security in cloud computing. *2020 10th Annual Computing and Communication Workshop and Conference (CCWC)*, Las Vegas, NV, USA, pp. 975–980. IEEE Xplore. doi: 10.1109/CCWC47524.2020.9031195.

Casola, V., Benedictis, A., De, Rak, M., & Villano, U. (2020). A methodology for automated penetration testing of cloud applications. *International Journal of Grid and Utility Computing*, *11*(2), 267–277.

Chinedu, P. U., Nwankwo, W., Aliu, D., Shaba, S. M., & Momoh, M. O. (2020). Cloud security concerns: Assessing the fears of service adoption. *Archive of Science and Technology*, *1*(2), 164–174.

Fatima, A., Khan, T. A., Abdellatif, T. M., Zulfiqar, S., Asif, M., Safi, W., Al Hamadi, H., & Al-Kassem, A. H. (2023). Impact and Research Challenges of Penetrating Testing and Vulnerability Assessment on Network Threat. *2023 International Conference on Business Analytics for Technology and Security (ICBATS)*, Dubai, United Arab Emirates, pp. 1–8. IEEE Xplore. doi: 10.1109/ICBATS57792.2023.10111168.

Ferry, E., O Raw, J., & Curran, K. (2015). Security evaluation of the OAuth 2.0 framework. *Information & Computer Security*, *23*(1), 73–101.

Gao, X., Liu, G., Xu, Z., Wang, H., Li, L., & Wang, X. (2020). Investigating security vulnerabilities in a hot data center with reduced cooling redundancy. *IEEE Transactions on Dependable and Secure Computing*, *19*(1), 208–226.

Haddon, D. A. E. (2021). Zero trust networks, the concepts, the strategies, and the reality. In *Strategy, Leadership, and AI in the Cyber Ecosystem* (pp. 195–216). Elsevier.

Hassanzadeh, Z., Biddle, R., & Marsen, S. (2021). User perception of data breaches. *IEEE Transactions on Professional Communication*, *64*(4), 374–389.

Hijji, M., & Alam, G. (2022). Cybersecurity awareness and training (CAT) framework for remote working employees. *Sensors*, 22(22), 8663.

Hock, F., & Kortiš, P. (2015). Commercial and open-source based Intrusion Detection System and Intrusion Prevention System (IDS/IPS) design for an IP networks. *2015 13th International Conference on Emerging ELearning Technologies and Applications (ICETA)*, 1–4.

Kashmar, N., Adda, M., & Atieh, M. (2020). From access control models to access control metamodels: A survey. *Advances in Information and Communication: Proceedings of the 2019 Future of Information and Communication Conference (FICC), Volume* 2, 892–911.

Kelley, P. (2023). *Evolution of Cyber Attacks and Their Economic Impact.* Authorea Preprints.

Kitchenham, B., Brereton, O. P., Budgen, D., Turner, M., Bailey, J., & Linkman, S. (2009). Systematic literature reviews in software engineering—A systematic literature review. *Information and Software Technology*, *51*(1), 7–15.

Kumar, G., Pandey, S. K., Varshney, N., Kumar, A., Kumar, M., & Singh, K. U. (2023). Cybersecurity Education: Understanding the knowledge gaps based on cyber security policy, challenge, and knowledge. *2023 IEEE 12th International Conference on Communication Systems and Network Technologies (CSNT)*, 735–741.

Kumar, R., & Goyal, R. (2020). Modeling continuous security: A conceptual model for automated DevSecOps using open-source software over cloud (ADOC). *Computers & Security*, *97*, 101967.

Maurer, T., & Hinck, G. (2020). *Cloud Security: A Primer for Policymakers.* Carnegie Endowment for International Peace.

Mehmood, A., Natgunanathan, I., Xiang, Y., Hua, G., & Guo, S. (2016). Protection of big data privacy. *IEEE Access*, *4*, 1821–1834.

Mishra, S., & Gochhait, S. (2023). Emerging Cybersecurity Attacks in the Era of Digital Transformation. *2023 7th International Conference on Intelligent Computing and Control Systems (ICICCS)*, 1442–1447.

Narasimhan, V. L. (2021). Using deep learning for assessing cybersecurity economic risks in virtual power plants. *2021 7th International Conference on Electrical Energy Systems (ICEES)*, 530–537.

Neto, N. N., Madnick, S., de Paula, A. M. G., & Malara Borges, N. (2021). A case study of the capital one data breach: Why didn't compliance requirements help prevent it? *Journal of Information System Security*, *17*(1).

Nightingale, A. (2009). A guide to systematic literature reviews. *Surgery (Oxford)*, *27*(9), 381–384.

Paiva, J. C., Leal, J. P., & Figueira, Á (2022). Automated assessment in computer science education: A state-of-the-art review. *ACM Transactions on Computing Education (TOCE*, *22*(3), 1–40.

Pedchenko, Y., Ivanchenko, Y., Ivanchenko, I., Lozova, I., Jancarczyk, D., & Sawicki, P. (2022). Analysis of modern cloud services to ensure cybersecurity. *Procedia Computer Science*, *207*, 110–117.

Ramakrishnan, R., Leethial, M., & Monisha, S. (2023). The future of cybersecurity and its potential threats. *International Journal for Research in Applied Science & Engineering Technology*, *11*(7).

Reddy, G. N., & Reddy, G. J. (2014). A study of cyber security challenges and its emerging trends on latest technologies. *International Journal of Engineering and Technology - UK*, *4*(1) ISSN: 2049-3444.

Santos, W. M., dos, Secoli, S. R., & Püschel, V. A. A. (2018). The Joanna Briggs Institute approach for systematic reviews. *Revista Latino-Americana De Enfermagem*, *26*, e3074.

Sharif, M. H. U., & Datta, R. (2020). Cloud data transfer and secure data storage. *International Journal of Engineering and Applied Sciences*, *7*(6), 11–15.

Singh, J. (2013). Critical appraisal skills programme. *Journal of Pharmacology and Pharmacotherapeutics*, *4*(1), 76–77.

Singh, K. K. V. V., & Gupta, H. (2016). A New Approach for the Security of VPN. *Proceedings of the Second International Conference on Information and Communication Technology for Competitive Strategies*, 1–5.

Srinivas, S., & Liang, H. (2022). Being digital to being vulnerable: Does digital transformation allure a data breach. *Journal of Electronic Business & Digital Economics*, *1*(1/2), 111–137.

Stafford, V. A. (2020). Zero trust architecture. *NIST Special Publication*, *800*, 207.

Sumadinata, W. S. (2023). Cybercrime and global security threats: A challenge in international law. *Russian Law Journal*, *11*(3), 438–444.

Taylor, A. R. E. (2021a). Future-proof: Bunkered data centres and the selling of ultra-secure cloud storage. *Journal of the Royal Anthropological Institute*, *27*(S1), 76–94.

Taylor, A. R. E. (2021b). Standing by for data loss: Failure, preparedness and the cloud. *Ephemera: Theory & Politics in Organization*, *21*(1).

Thilakarathne, N. N., & Wickramaaarachchi, D. (2020). Improved hierarchical role based access control model for cloud computing, *International Research Conference on Smart Computing and Systems Engineering, ArXiv Preprint ArXiv:2011.07764*.

Torkura, K. A., Sukmana, M. I. H., Cheng, F., & Meinel, C. (2021). Continuous auditing and threat detection in multi-cloud infrastructure. *Computers & Security*, *102*, 102124.

Tsegaye, T., & Flowerday, S. (2020). A Clark-Wilson and ANSI role-based access control model. *Information & Computer Security*, *28*(3), 373–395.

Upadhyay, D., & Sampalli, S. (2020). SCADA (supervisory control and data acquisition) systems: Vulnerability assessment and security recommendations. *Computers & Security*, *89*, 101666.

Vegesna, V. V. (2023). Utilising VAPT technologies (Vulnerability assessment & penetration testing) as a method for actively preventing cyberattacks. *International Journal of Management, Technology and Engineering*, *12*.

Wu, C., & Li, W. (2021). Enhancing intrusion detection with feature selection and neural network. *International Journal of Intelligent Systems*, *36*(7), 3087–3105.

Xiao, Y., & Watson, M. (2019). Guidance on conducting a systematic literature review. *Journal of Planning Education and Research*, *39*(1), 93–112.

Yayla, A. A., & Hu, Q. (2011). The impact of information security events on the stock value of firms: The effect of contingency factors. *Journal of Information Technology*, *26*, 60–77.

Yeoh, W., Liu, M., Shore, M., & Jiang, F. (2023). Zero trust cybersecurity: Critical success factors and a maturity assessment framework. *Computers & Security*, *133*, 103412.

Zeng, X., Garg, S., Barika, M., Bista, S., Puthal, D., Zomaya, A. Y., & Ranjan, R. (2020). Detection of SLA violation for big data analytics applications in cloud. *IEEE Transactions on Computers*, *70*(5), 746–758.

Zgureanu, A. (2021). Backup and recovery strategies and their role in business continuity. In The Collection (pp. 285–293).

Chapter 2

A new framework for agile cybersecurity risk management

Integrating continuous adaptation and real-time threat intelligence (ACSRM-ICTI)

Mounia Zaydi, Yassine Maleh, and Youness Khourdifi

2.1 INTRODUCTION

To address the ever-evolving and complex challenges posed by the rapidly advancing digital landscape, organizations are actively pursuing a risk management approach with the dynamic and adaptable characteristics of agility. Agile risk management serves as an innovative and effective approach that empowers organizations to effectively identify, assess, and proactively handle risks while seamlessly adjusting to changes within the information technology (IT) environment and the intricate risk landscape (see Figure 2.1 for a visual representation of the evolution of the digital landscape and its associated risks).

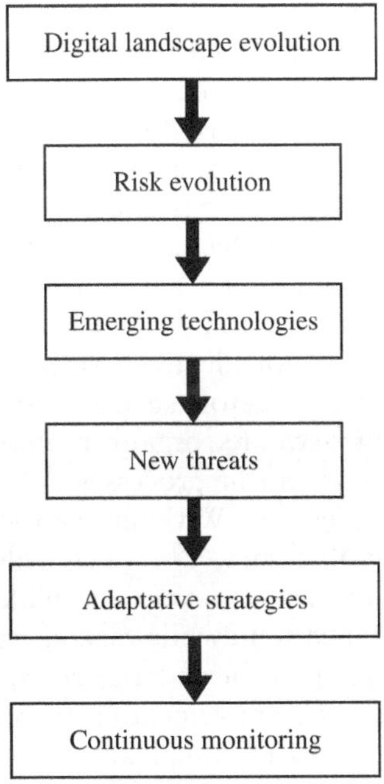

Figure 2.1 Evolution of the digital landscape and associated risks.

DOI: 10.1201/9781003478676-2

However, despite its immense potential, the concept of agile risk management has lacked a comprehensive framework or guideline that outlines the seamless integration of agility into risk management practices. Recognizing this critical gap, this groundbreaking research is committed to providing an unprecedented, comprehensive, and remarkably practical framework for the domain of agile cybersecurity risk management (CSRM). By equipping organizations with this indispensable tool, unparalleled protection in the face of constantly evolving cyber threats and intricate IT environments will be ensured [1–3].

Numerous CSRM frameworks and standards have been published to assist organizations in efficiently and systematically managing risks. These frameworks aim to provide comprehensive guidance and support in tackling cybersecurity threats. Among these frameworks, the National Institute of Standards and Technology (NIST) framework stands out as the most popular and comprehensive. NIST's framework emphasizes the importance of a risk management process that encompasses risk assessment, risk mitigation, and continuous assessment through a feedback loop. In addition to NIST, other notable IT governance and auditing standards and guidelines such as COBIT and ISACA also offer detailed risk management recommendations. The key concept underlying these frameworks and standards is the integration of risk management into an organization's overall management and governance. By seamlessly embedding risk management practices, organizations can effectively address cybersecurity risks within the broader context of their operations. However, the ever-evolving landscape of cyber threats and the increasing complexity of IT environments pose significant challenges for organizations in their efforts to manage cybersecurity risks successfully. One of the common challenges reported by organizations is the difficulty in identifying and assessing risks accurately. With cyber threats becoming more sophisticated and diverse, organizations struggle to keep pace and accurately identify potential risks. Additionally, the assessment of risk impact on the business can be complex, as it requires a comprehensive understanding of the organization's operations, assets, and dependencies. Determining the appropriate response and treatment for identified risks is also a challenging task, as it involves weighing different factors such as cost, feasibility, and potential impact on operations. Furthermore, organizations often face the challenge of reviewing their risk management processes and evaluating the effectiveness of their risk mitigation efforts. Without historical data and insights into previous risk management outcomes, it becomes arduous to assess the overall effectiveness of risk management practices and make informed decisions for future improvements. In conclusion, while various frameworks and standards provide guidance for managing cybersecurity risks, organizations encounter several challenges in effectively implementing these practices. The dynamic nature of cyber threats and the complexity of IT environments require organizations to continually adapt their risk management strategies to stay ahead. Addressing challenges such as risk identification, assessment, response determination, and process evaluation is crucial for organizations to enhance their cybersecurity posture and safeguard their valuable assets [4–7].

This chapter is structured to provide a comprehensive understanding of the principles and necessity of an agile CSRM framework. It starts with an exploration of agile methodologies in CSRM, highlighting how these approaches can enhance responsiveness and effectiveness in an ever-evolving digital landscape. Subsequently, we delve into the integration of real-time threat intelligence (RTTI), examining its crucial importance for proactively anticipating and responding to threats. The following section is dedicated to continuous adaptation in the face of an evolving threat landscape, detailing strategies for maintaining a dynamic and resilient security posture. Finally, we introduce our proposed contribution: a proactive framework incorporating RTTI, designed to bolster CSRM through advanced technologies and systematic threat data integration. This framework aims to transform risk management practices into more agile and informed processes, thereby offering enhanced protection in the global digital environment. Continuing from this foundation, the text explores the transformative outcomes anticipated from deploying such a dynamic CSRM framework. This proactive approach to cybersecurity is poised to revolutionize traditional reactive security measures by enabling organizations to not only respond to threats more swiftly but also anticipate them, significantly reducing the incidence and impact of security breaches. Furthermore, the discourse concludes by contemplating the future of agile cybersecurity frameworks, advocating for the integration of emerging technologies such as artificial intelligence (AI) and machine learning. This forward-looking perspective is aimed at further enhancing the collection and analysis of real-time threat data, thus continually refining decision-making processes and ensuring that cybersecurity strategies evolve in alignment with both technological advancements and emerging threat landscapes.

2.1.1 Background

Risk management is a well-documented discipline derived from the actuarial, insurance, and decision sciences. As with predictive/preventative sciences such as quality and reliability engineering, it has proven difficult to apply the disciplines of risk management across a broad spectrum of security domains. As a result, risk management in cybersecurity has primarily evolved through compliance regulations, which are external pressures that force a domain of control onto practices of required adherence. The compliance regulations establish a set of control objectives but do not enforce practices that mandate measurable reduction of risk. This differs from organizations whose passenger seat safety must adhere to Federal Motor Vehicle Safety Standards (FMVSS) crashworthiness standards, as failure to do so results in product recall and penalty of fines. The absence of enforced risk reduction practices has led to overly qualitative risk analysis often done by manual spreadsheets and intuition. This has been cost prohibitive for smaller organizations who cannot afford headcount of policy and risk analysts. It has also led to disconnect with technology practitioners who are tasked with security enforcement as they do not see credible intelligence translated into action that makes their ecosystem more defensible. The intelligence and decision support aspect of risk management is the least evolved in the cybersecurity domain and presents an opportunity to create practices that are proactive. Mitigation intelligence has too often been the result of post-incident analysis and therefore reactive change in controls. Published incident analysis often results in global signatures and public-facing indicators of compromise which have forced a landscape change that benefits the adversary. The opportunity here is to create change in control that is focused and quiet with least impact to normal business serving as deterrence to adversary tactics. To address the challenges of risk management in cybersecurity, it is essential to enhance and expand upon the existing practices and methodologies. By amplifying the focus on compliance regulations, organizations can establish more comprehensive control objectives that promote measurable reduction of risk. This will ensure that risk reduction practices are enforced consistently and effectively across different security domains. Moreover, it is crucial to employ advanced technologies and tools that can automate the risk analysis process, reducing dependency on manual spreadsheets and intuition. This will make risk management more cost-effective for smaller organizations and enable them to allocate their resources more efficiently [8–10]. To bridge the gap between technology practitioners and risk analysts, it is important to enhance the translation of credible intelligence into actionable measures. This can be achieved by developing sophisticated decision support systems that provide real-time insights and guidance for security enforcement. By integrating intelligence-driven strategies, organizations can proactively identify potential threats and vulnerabilities, allowing them to implement targeted and effective control measures. Additionally, there is a need to shift from reactive change in controls to a more proactive approach. Instead of relying solely on post-incident analysis for mitigation intelligence, organizations should invest in predictive analytics and threat intelligence platforms. This will enable them to anticipate and prevent cyber incidents, reducing the impact and damage caused by attacks [11, 12]. By leveraging these technologies, organizations can create a landscape that is more resilient and less susceptible to adversary tactics. Furthermore, it is crucial to develop discreet and focused changes in control that minimize disruptions to normal business operations while serving as a strong deterrent to adversaries. In conclusion, the evolution of risk management in cybersecurity requires a comprehensive transformation of practices and methodologies. By enhancing compliance regulations, leveraging advanced technologies, fostering collaboration between technology practitioners and risk analysts, and adopting proactive approaches, organizations can establish a more robust and effective risk management framework (RMF). This will ultimately contribute to a safer and more secure cyber ecosystem, protecting both organizations and individuals from potential threats and vulnerabilities [13, 14].

2.1.2 Objectives

This chapter aims to outline our strategic goals for advancing CSRM that are as follows:

- **Develop an Effective and Realistic Approach to CSRM**: Adapt this approach to both security threats and changes in the business environment to ensure a dynamic and relevant response.
- **Create a Security RMF Based on International Best Practices and Standards**: Enable a consistent approach to managing information security risks.
- **Study Risk Acceptance Levels and Develop Appropriate Tools:** Equip business sectors with the tools to make informed and consistent decisions regarding the acceptance of security risks.
- **Enable Informed Decisions on Security Risk Levels and Risk Treatment Options:** Improve the availability, accuracy, and relevance of security risk information to businesses.
- **Demonstrate Measurable Improvement in Security Risk Management Practices**: Reduce security risks through tangible and verifiable improvements in risk management.

2.1.3 Scope

In the digitized environment, information systems are being constantly threatened by the regular occurrence of various security incidents and persistent penetration attempts. This security issue is made more complex with the modern trend of outsourcing where organizations expose their sensitive information through networks to customers, suppliers, and other parties that need to use that information. Often, limited budgets and resources have encouraged organizations to accept the current level of security as "good enough." When losses from security breaches only become serious enough to be noticed, they will consider increasing security. However, this normally reactive process occurs too late with the damage already done, and in the end, the cost of increasing security at that point and the loss far exceeds the cost of proactively increasing security. CSRM is the identification, protection, detection, response, and recovery of system security to decrease the likelihood of security breaches and reduce damage and cost of those breaches. Traditional risk management often involves trade-offs between increased security and cost and seeks to find an acceptable level of risk [8, 14]. Ongoing risk management sees that system changes are considered, but both seek to define a static set of security requirements to be met with a security posture and focus on operating within a risk tolerance and security gains are often minimal because they are difficult to quantify at the executive management level where resource allocation decisions are made. These traditional and often siloed approaches of functional security and risk management change often leave security as an afterthought, and the current dynamic and global security threats have revealed an urgent need for CSRM that is fully integrated into system and business life cycles during system operation and evolving business and mission needs [15–17].

2.2 AGILE METHODOLOGIES IN CYBERSECURITY RISK MANAGEMENT

Cybersecurity has always been a significant challenge for businesses and industries throughout the planet, with actual cyberattack and serious compromises in systems growing over the last decade. In this chapter, it is our aspiration to indicate the existing theories and methods which have been used to treat and talk about cybersecurity dangers, and then to indicate that there's a higher manner. This can be finished by figuring out a few basic assumptions and developments in cyber threats and cyber-protection, and then by drawing on analogies from other forms of threat and hazard management to encourage changes in approach. The default method to cybersecurity hazard management that's employed by most organizations at the present time carefully resembles methods to first-rate management in many specific industries. For a comparative overview of traditional and agile methodologies, refer to Figure 2.2.

By focusing on regulatory/legislative compliance and risk evaluation using annual, bi-annual, and quarterly critiques, it's a batch process that tries to reduce risk by reaching levels of quality and security that are deemed acceptable. Simulation of the methods used in attempts to change business dynamics which were copied from other industries suggests that they're uncompetitive and leave essential strategic goals unaddressed. This chapter will argue that a better method for cybersecurity hazard management lies in accepting

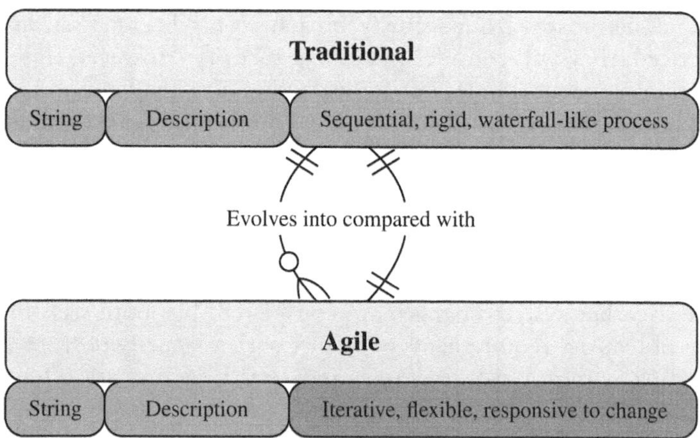

Figure 2.2 Comparison of traditional and agile methodologies.

ever-present change in cyber threats and strategically adopting higher hazard with a purpose to get greater results. To guide this argument, a path toward the desired state of continuous adaptation can be identified, with clear milestones and tactics [18–20].

2.2.1 Overview of agile methodologies

In the early 2000s, the term "agile" was defined in the Agile Manifesto. This was a reaction to various lightweight software development methods of the time, based on codifying some of the new, less documentation-centric methodologies. A core tenet of the agile movement is that the better way to develop software is to do it iteratively and that it is better to build something small than to plan to build something big. The manifesto included 12 principles that underpin the four defined values. The Agile Manifesto is individuals and interactions over processes and tools, working software over comprehensive documentation, customer collaboration over contract negotiation, responding to change over following a plan. A few lightweight development methods evolved in reaction to the bureaucracy inherent in traditional methodology. These included Scrum (1993), which took place in the industry of complex products, systems, and new technologies; Dynamic Systems Development Method (DSDM) (1994), a consortium of vendors and experts in Europe formed to unify the various lightweight methods being used; and Extreme Programming (XP) (1996), a method which embodies the new philosophy in any methodology that lightweight is better. The early methods focused just on better, more lightweight ways of developing software. XP was an explicit attempt to define a method that did not challenge the claimed problems of the waterfall method (inability to deal with change, high defect rates, late, over-budget releases) but rather provide a set of tools to achieve them. Only after experience had shown that these methods were in fact a better way to develop software did the agile movement define itself with the publication of the Agile Manifesto. XP tried continually to base its practices on the results of controlled experiments [21, 22].

2.2.2 Benefits of agile methodologies in CSRM

In an analysis of agile practices in a regulated industry, authors of [23] recognized that to prevent compromise of regulations, controls, and best practices, security should not be considered an external quality attribute. Instead, it should be an inherent property of the system being developed. If regulations and best practices are treated as requirements, security tasks will be integrated into the standard agile planning and will evolve along with the system. This provides a significant advantage over systems where security is considered after the fact or IT systems where security tasks are often avoided or postponed, seeking to achieve "good enough" security. Finally, work on the development of a safety system based on genetic algorithms demonstrates that the ability to respond quickly to changes in the risk environment can be

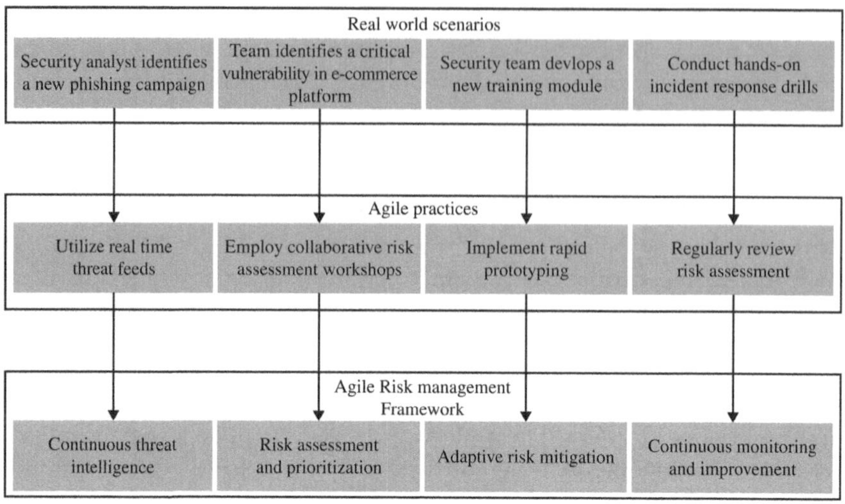

Figure 2.3 Examples of agile practices in risk management.

highly beneficial. To achieve this, the system itself must evolve in alignment with the risk management process [24, 25]. See Figure 2.3 for examples of agile practices in risk management.

The benefits of agile methodologies are well recognized in the software and project management domains. Agile methodologies encourage iterative and adaptive development, which can effectively support managing cybersecurity risks. In situations where controls and countermeasures may be uncertain and rapidly outdated, traditional risk management methodologies attempt to govern the process of identifying, assessing, and mitigating risks. However, these processes can be highly disjointed, yielding artifacts that quickly become stale. They effectively neglect risk treatments where the procedure to mitigate a risk may be uncertain or would incur great expense. By contrast, agile risk management integrates the risk management process into agile development and aims to reduce risks by iteration [26].

2.2.3 Challenges and considerations

Due to the unique nature of cybersecurity threats, the traditional agile methodologies used in IT and business often cannot be directly applied to managing risk in cybersecurity. Occasionally, companies might implement software development agile methodologies into risk management and overcomplicate the process, possibly introducing new vulnerabilities as a result. Additionally, CSRM is a high stake, high-pressure process. Acting in an agile manner often requires fast-paced changes that large, slow-moving, or risk-averse organizations might struggle with. A mistake in an agile risk management approach can lead to vulnerability or a successful attack. Successful attacks can be catastrophic for organizations and failure in risk management due to misplaced agility can even result in regulatory fines or legal action. For these reasons, there is considerable debate on whether agile methodologies are a good choice for CSRM. This research aims to develop a framework incorporating Real Options Theory and a systematic approach based upon the NIST SP800-39 RMF that will allow an organization to make risk-effective risk management decisions between agile and non-agile approaches [27–29].

2.3 REAL-TIME THREAT INTELLIGENCE INTEGRATION

The continuous evolution of information systems and the increase of attack severity have created a need for RTTI to be integrated into modern risk management strategies. The very nature of cyber-attacks is dynamic and can cause substantial damage before countermeasures can be put in place. Thus, the ability to prevent attacks from occurring in real time is more desirable than being able to recover from them after the fact. Traditional risk management typically involves static security controls and post-analysis of events. These controls and analyses are generally based on past experiences and designed to prevent re-occurrences of specific incidents. This form of risk management is not effective against modern threats, is a poor use of resources, and can create a false sense of security. The knowledge obtained from RTTI is designed to bridge informational gaps, influence smarter security controls, and provide a clear understanding of current security postures. This form of risk management is proactive in nature and can provide continuous adaptation alongside the attacks it is trying to prevent [30, 31] (Figure 2.4).

2.3.1 Understanding real-time threat intelligence

Ideally, those defensive actions are chosen to contain or remove threats that are targeting the organization preserving or enhancing its security posture. To enable that choice and measure its impact, defenders need to be able to

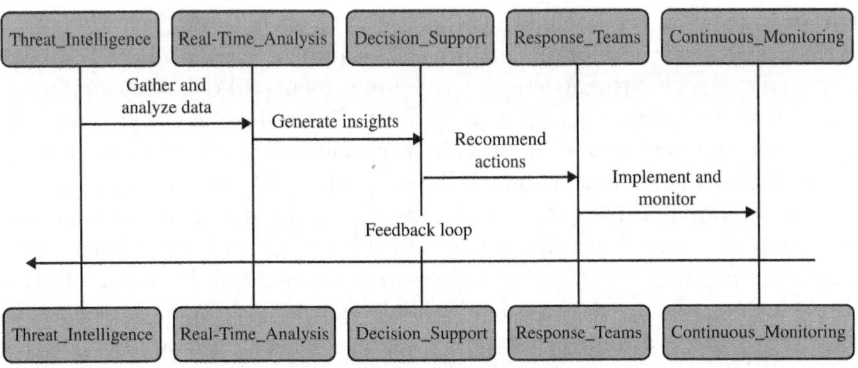

Figure 2.4 The process flow of real-time threat intelligence.

represent the state of the organization's security posture, the threat's actions against it, and the probable effects of their own actions in a formal way. Then they can relate these representations to one another, reason about the problem space, and determine the best course of action. This kind of control and analysis is based on having a clear and timely understanding of the security posture of the organization and the threats it faces; these are areas where cybersecurity risk management has traditionally fallen short and where RTTI has a unique opportunity to make a difference in security outcomes [32–34].

Combating a more sophisticated and persistent adversary has driven the need for cybersecurity approaches that are more dynamic and quicker to adapt intelligence-driven approaches to cyber defense, which is one way of bolstering the ability of organizations to keep up with the rapid changes in the cyber environment. A current deficiency with existing intelligence-driven approaches to cyber defense is that they lack a foundation for sharing and utilizing common understanding (intelligence) at a fast enough pace to enable the required adaptation. Although intelligence is an essential building block to decision-making in cyber defense, too often its impact is isolated to specific actions like blocking an IP address, isolating a compromised system, or taking a known indicator of compromise and seeking it across the enterprise. These actions are important, but in aggregate, they rarely change the overall security posture of the organization [35, 36].

2.3.2 Integration of real-time threat intelligence in the framework

To achieve these high levels of valued intelligence, an organization must first understand the cost and the common barriers associated with traditional intelligence. The main driver behind the framework development was to undeniably affect the agility management and solution within the intelligence-driven security space. In doing so, the first required adaptation is the recognition that the value level of intelligence should directly affect the security posture. This will allow an organization much more cost-effective prioritization and effective utilization of high-valued intelligence. The second required adaptation is the acceptance that not all security risks can be averted and the best solution against specific threats may be accepting calculated risk and using threat acceptance decisions. This must be supported with flexible risk management functions, and the framework provides the solution through the alignment of decision-based requirements on intelligence value and risk posture [37, 38].

Organizational intelligence, including the combination of internal and subscription external intelligence, is somewhat more valuable, though often requires customization to avert information overload and the use of effective solutions specific to the organization's security risk management needs. This, however, is a higher cost, lower agility solution, as the intelligence is often too late, and the solutions ineffective. These barriers become non-issues when considering that only through achieving the intelligence-value level of

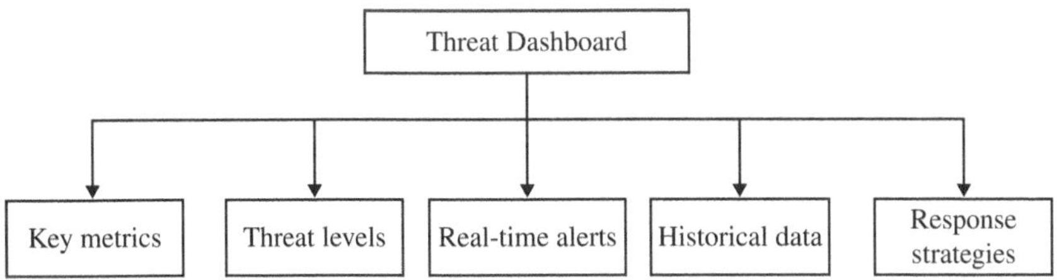

Figure 2.5 Example threat intelligence dashboard.

adversary intelligence can an organization fully understand their adversaries and effectively defend against them. Today's adversaries use varying tactics and develop specific purpose attacks against targeted organizations. Only with intelligence at this level can an organization detect such threats, prepare to defend, or launch a counterattack [39, 40]. An example of a threat intelligence dashboard can be seen in Figure 2.5.

Currently, there is a lack of focus on threat intelligence due to the low priority it is often given and the negative impact it can have on rapid business change and agility. Traditional intelligence is describing general purpose security information, much of it outdated and wholly ineffective when it comes down to solving the security problems of today. This limitation can be seen in the fact that many organizations solely rely on freely available intelligence resources, the same ones being utilized by countless others. As a result, the information is of low value and the recommended security solutions are too generic and lack effectiveness [41, 42].

2.3.3 Enhancing detection, analysis, and mitigation

Once threat intelligence has been generated, it needs to be used in changing the environment to mitigate threats and reinforce defenses. This is a highly complex and varied process that can involve changes to data and network configurations, system hardening, and in some cases changes to hardware and system architecture. The key to success is being able to exert a high level of control over a system to make the right changes in an efficient and low-risk manner. This can involve using the threat intelligence to automatically make changes to the environment but often it is more effective for an analyst to manually make changes while being guided by the threat intelligence [43–45].

New approaches and tools are being developed in data collection and analysis to facilitate the generation of threat intelligence from the vast amount of varied data present in IT environments. These new approaches often center around the concept of situational awareness where an accurate and complete understanding of the environment is built by correlating different sources of data into a meaningful representation of what is happening in an environment and identifying changes to this representation as indicators of potential threats. The key is to correlate complex and varied data in a way that makes it easily understandable and actionable for an analyst [42, 46, 47].

The traditional approach to detection, analysis, and mitigation of threats relies on running various automated tools to detect changes in the environment and sending the data to analysis tools or storage in databases. This approach is effective only at finding known threats and does not scale to handle the vast amount of data and potential threats present in the dynamic and complex IT environments of most organizations. The key to effectively detecting, analyzing, and mitigating threats is to leverage the vast amount of new and varied data present in IT environments to generate intelligence about potential threats and use this intelligence to drive change in the environment to block detected threats and reinforce defenses in areas where vulnerabilities have been exploited [48, 49].

2.4 CONTINUOUS ADAPTATION TO THE CHANGING THREAT LANDSCAPE

It is generally agreed upon in the cybersecurity community that the landscape of threats is continuously changing. The understanding of the threat points that cyber adversaries use to attack systems is increasing and the adversaries are continually changing and improving their tactics, in response to the increased understanding of the threats they face. One basic example of this is the increasing prevalence of malware that specifically targets the boot record or the master boot record of a system. Malware of this nature is particularly difficult to remove and has a high likelihood of re-infecting the system. This type of malware is in response to the higher resilience and efficacy in removing traditional malware from systems. Another example is the increased prevalence of attacks on virtual infrastructure. As companies increasingly virtualize their infrastructure, cyber adversaries are taking note and developing tools and methods to attack the virtual infrastructure. As the prevalence of both types of attacks increases, it is clear evidence that the adversaries are adapting to the changing nature of systems and their attack points. These examples serve to illustrate the common adage in the cybersecurity community that "adversaries are always one step ahead." (The continuous adaptation cycle is illustrated in Figure 2.6.) Because the threat landscape is continually changing and that adversaries are constantly adapting their attacks to new and existing systems, it is necessary for a CSRM framework to include methods to continuously adapt to the changing threat landscape [50, 51].

2.4.1 Importance of continuous adaptation

Any effective cybersecurity risk management process entails a comprehensive understanding of the threats, vulnerabilities, and potential adverse impacts. As mentioned in the previous section, the threat landscape is constantly evolving and becoming more complex. Traditional risk management following the system of the NIST Risk Management Framework was not designed to consider

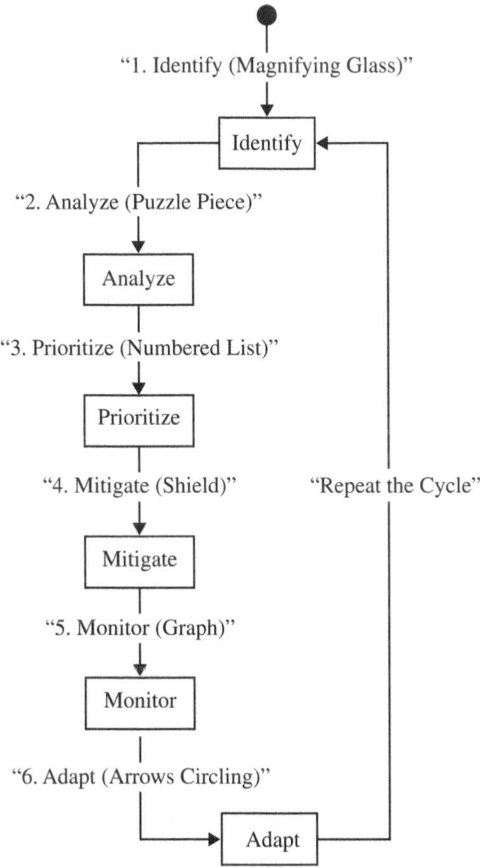

Figure 2.6 Continuous adaptation cycle in cybersecurity risk management.

the dynamics of the threats it aimed to manage and thus was based on a snapshot in time of the threat environment. For a general risk management process, this would not be problematic. However, for Enterprise Security Plan (ESP) and its duplication of the RMF, it means that the various security plans, security controls, and allocated resources will become rapidly outdated, and the player will be left with a false sense of security or potential concern due to the difference in environment from what was anticipated. This obsolescence of information can lead to wasted resources, as the execution of security controls may no longer be appropriate to the changed threat environment, and ineffective decision-making as mentioned in the OODA (Observe, Orient, Decide, and Act Loop) loop due to the mismatch of the environment in the decision maker's mind and the real environment. Cybersecurity has the added difficulty that the bidirectional dependencies between the actions of the player and the changes in the threat environment can cause rapid and radical changes in the environment. This is termed the Red Queen effect after Lewis Carroll's *Through the Looking*

Glass, though the term has also been used to describe an arms race or competitive systems balance in evolutionary biology. Continuing to play the same game, namely the defense of information with attackers, essentially moves no one on the game board but can radically deplete resources as competition is played out, eventually leading to a decreased quality in the service. This can be seen in modern business where both legitimate organizations and adversaries have engaged in Distributed Denial of Service (DDoS) attacks. With the intangible nature of the cyber environment and global information society that more resembles an interconnected ecosystem, a purely static model of security plan with enacted security controls just does not suffice [7, 52].

2.4.2 Strategies for continuous adaptation

In documenting the thorough strategies needed to adapt, Cybersecurity Agile Risk Management (CARM) is breaking new ground in the cybersecurity risk management field, where previous research has glossed over the necessity for ongoing change. The reader is confronted with the unpleasant truth that the threat environment is never constant, and even the best risk management practices are completely ineffective if they remain static. The strategies are derived from the capabilities provided by the CARM framework's core components of dynamic risk assessment and intelligent mitigation. Changes to risk assessment involve tracking the evolution of threats and vulnerabilities to better predict likely outcomes, such as attacks on systems or information assets. This involves correlating RTTI about adversaries and their objectives with an understanding of the potential for exploiting vulnerabilities in different systems to determine the likelihood of successful attacks. Recommended methods include tracking data such as Common Vulnerability Scoring System (CVSS) scores and expert systems used to analyze historical attack and vulnerability data. Changes to intelligent mitigation mean adapting decisions about accepting, avoiding, mitigating, and transferring risk. This means continually reassessing mitigation measures and their effectiveness as the cost and/or risk of the impact of a successful attack on a protected asset change. This decision process can be visualized using decision trees containing predefined mitigation strategies and calculated probabilities of success, allowing easier adaptation in the event of change. An example of a decision rule would be "if the cost of its failure is too high, abandon cloud storage of X data." This contrasts with the current common practice of making ad hoc mitigation decisions based on static risk assessments [53, 54].

2.4.3 Leveraging cutting-edge technologies

A model-based reasoning system is an example of a more advanced decision-making technology. This has potential application in assessing probable consequences of system deployment or reconfiguration. The possible actions are represented as a set of transitions between system states, and the effect of each

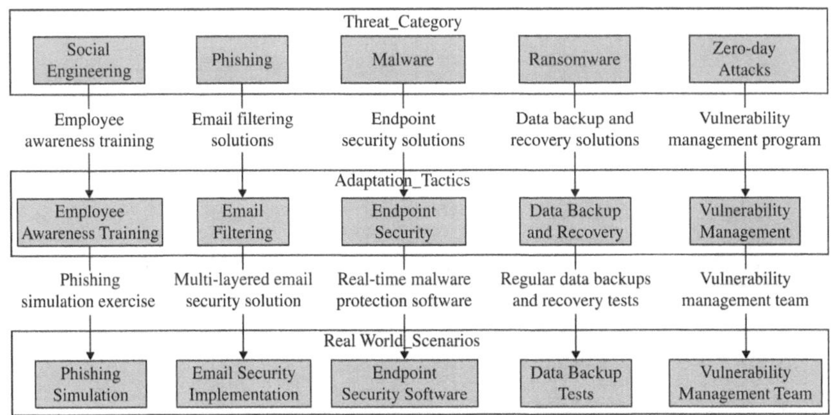

Figure 2.7 **Examples of emerging threat adaptation tactics.**

action is evaluated in terms of changes to the probability of system states. If the system can be thought of as a utility maximizer, it will then choose the action that is expected to have the best consequences. This approach reduces the amount of real-time decision-making required at critical periods, thus reducing human error probability by skipping or delaying tough decisions and allowing the planning of system reconfigurations in low threat periods [55, 56].

A keyway to enable continuous adaptation is through the development and implementation of technologies in CSRM that automate and enhance decision-making. Decision-making in uncertain, high threat environments is a central concern to the security community. At one extreme, one could imagine a system that makes all possible decisions with respect to the threat environment and its possible consequences. In the absence of complete, perfectly correlated information about the threat environment and the efficacy of the system's possible reactions, there is a certain probability that the system will choose a course of action that is suboptimal, possibly leaving the system and the resources it is trying to protect in a less favorable state [52, 57] (Figure 2.7).

2.5 PROPOSED CONTRIBUTION: A PROACTIVE FRAMEWORK INTEGRATING REAL-TIME THREAT INTELLIGENCE

This contribution introduces a new framework designed to enhance CSRM through the integration of agile methodologies and RTTI. The framework is structured to allow for continuous adaptation to the changing threat landscape and integrates cutting-edge technologies to improve the detection, analysis, and mitigation of cybersecurity risks. Taking a top-down approach, the RMF begins with a governance tier designed to define metrics and overall risk strategy. This leads to the action tier which implements real-time security analysis and then develops and assigns security-relevant tasks to better mitigate risks. Finally, the framework has tiers for issue resolution and

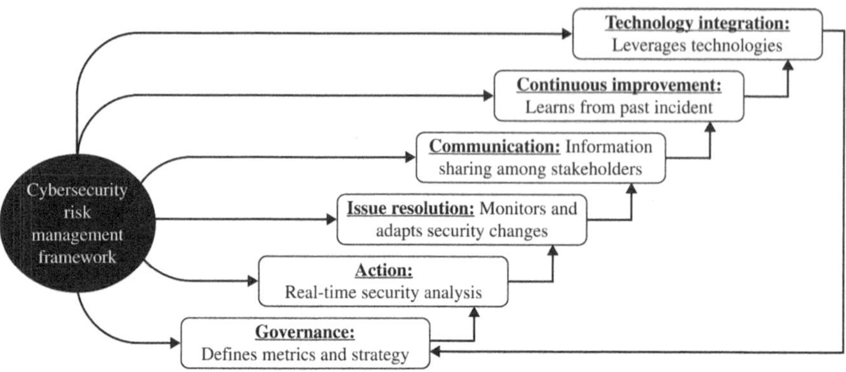

Figure 2.8 Global architecture of the proposed risk management framework.

communication, which continuously monitor and adapt the security changes implemented into the overall framework. The goal of the RMF is to enable an organization to leverage a repeatable, cost-effective process of implementing and/or improving security using a framework such as this; the global architecture of the proposed framework is detailed in Figure 2.8.

2.5.1 Key components of the framework

The key paradigm shift in our framework is the linkage of risk management with real-time system state awareness for security, support of traditional risk management evaluation, and continuous system monitoring and adaptation. In the event of a data log, system state awareness, and experiential evidence, is used for evaluating security risk, assigning countermeasures, and understanding the effects of the risk mitigation in each operational system to ensure that the system is operating within acceptable risk tolerances. Instead of periodic security/risk assessment which occurs in the current RM and IA processes, system state awareness data is continually used to reassess security risks and evaluate the effectiveness of risk treatment options while considering the current state of the system. This focuses risk management into a continuous and active process, assessing and treating risks as they develop and occur in the system in contrast to the current culture of periodic assessment and treatment of past or potential future risks. The primary enabler to this in the framework is the use of computational simulation of system models to automate the analysis and assess many possible solutions in complex operating systems. This minimizes the human in the loop analysis that makes assessing and acting on system vulnerabilities cost prohibitive. The comparison of action selection and anticipated effects is a decision process known as intelligence which is a required function for any sort of automation or robotic system. For an autonomous or self-directing system, computational intelligence about the system and its environment is required to enable the system to make informed decisions. Intelligence in the framework is the most effective means to provide decision support for evaluating risk treatment options and understanding the effects of those options on the system. Simulation and intelligence provide a way to bring actionable results and metrics to the traditionally qualitative and procedural processes of RM and IA. Simulation run data is akin to test data gathered from an experiment. It is recorded information about the action and the state of the system while the data is used to compare to expected results and identify discrepancies. In the framework, simulation run data is used to compare the current and expected system states and identify security risks that may affect the system from the changes to the action. The use of run data and the simulation itself provide a safe and low-cost environment to test and identify vulnerabilities and the effects of security measures to prevent unwanted changes to the actual operating system. Simulation is also a means to proactively identify risk in future system changes and compare the risk of different possible futures to support informed decisions on the best course of action. Simulation and intelligence together provide a measured and scientific approach to risk management in understanding and taking action to control system risk and evaluate the effects of those actions. Simulation and intelligence are a complex and advanced application and the use of these methods in RM and IA is an innovative and high-impact solution to systemic problems in assuring and evaluating security posture in an ever more dynamic and hostile IT landscape.

2.5.1.1 Agile planning and execution

The Plan-Do-Check-Act cycle from quality management provides a proven framework for executing iterative risk management plans. This is the foundation of agile risk management implicit in the various agile software development methodologies. In the planning phase, the mitigation strategy with the highest priority risks is planned and then the strategy is executed. The results are then evaluated in the context of the current knowledge about risk and its potential consequences. These results should feed back to the risk assessment phase and result in reassessment of risk probability and impact. This should then lead to adaptation of the risk mitigation plan for those risks, or identification of other risk events where the knowledge gained indicates a need to change the plan. At any given time of the cycle, there is a plan with an associated probability of success which is subject to continuous assessment and adaptation. An example of executing this cycle might be to compare two alternative security system configurations using the means available in the second to assess and adapt the system of the first [58].

Planning is among the most fundamental activities in risk management involving both conscious and unconscious processes. Underlying risk management plans, there are a set of assumptions and heuristics about deterrents to future risk (or facilitators to risk reduction) and the probability of future risk events. The assumptions and heuristics may be partially articulated in a loosely connected series of if-then scenarios and plans for risk mitigation. However, research in decision science indicates that people are notoriously bad at assessing the probability of future events. That coupled with the dynamics of complex systems means that risk assumptions are often invalid when the risk eventuated, and the initial risk assessments were made. Therefore, plan adaptation is vital to managing risk in complex environments.

- **Iterative Risk Assessment:** Regularly update and refine risk assessments to reflect new data, evolving threat vectors, and changes in organizational assets.
- **Adaptive Risk Mitigation Strategies:** Develop flexible mitigation strategies that can be quickly adjusted or scaled based on real-time risk analysis and threat intelligence.

Successful risk management is an iterative process which must rapidly react to changes in organizational assets, the threat environment, and risk posture. Traditional approaches to risk management are often static and inflexible, failing to accommodate the rapidly changing nature of cyber threats and the environments in which they occur. By treating the initial risk assessment as a baseline measure of risk, the differences between current and predicted future assessments represent changes in risk levels due to the threats and vulnerabilities that have been realized and changes in the internal/external risk environment. By calculating the rate of change in risk levels over given time intervals, it is possible to identify emerging threats and the effect of changes to assets on the overall level of risk. The key challenge here is to identify whether the changes in risk levels are acceptable, or whether action should be taken to mitigate/reduce the risk to acceptable levels. Through identifying whether risk levels have moved in or out of defined risk acceptance thresholds, this method provides a means of risk-based decision-making for change and the prioritization of resources to areas where risk levels are not acceptable. Simulation of alternative actions and their effects on the rate of risk levels can also provide a decision support mechanism for the identification of best courses of action to increase or reduce risk in given areas. This general approach adequately describes many aspects of iterative risk management and can be adapted to provide intelligence-driven decision support and action identification in specific areas of risk.

2.5.1.2 Incident resolution

This pivotal phase in cybersecurity incident management is essential for effectively addressing, mitigating, and preventing the recurrence of security breaches. It encompasses several key components aimed at resolving incidents with minimal operational disruption. Initially, incidents are identified and classified through real-time analysis based on their nature and severity, which helps prioritize responses. This is followed by a swift impact assessment to evaluate potential business effects, such as data loss and service interruptions. Containment measures are then implemented to isolate and limit the spread of threats, complemented by efforts to eradicate the threat source through malware removal and vulnerability repairs. Subsequently,

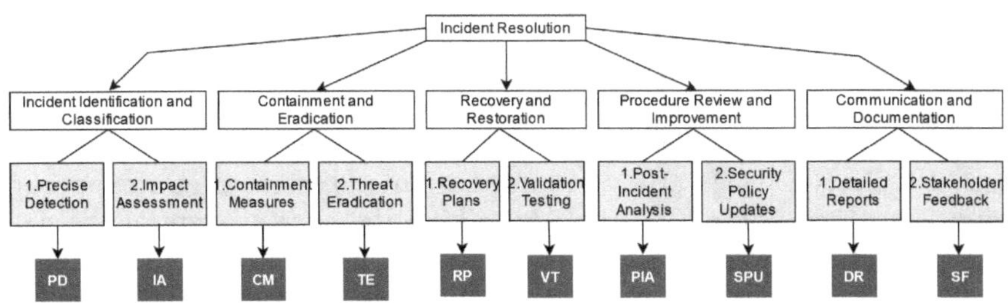

Figure 2.9 Incident resolution process flowchart.

recovery plans are developed and executed to restore affected systems, minimize downtime, and ensure system integrity. Rigorous validation testing is performed before systems go back online. The phase also involves a thorough review and improvement process, including detailed post-incident analysis to understand root causes and update security policies and procedures based on the lessons learned. Furthermore, comprehensive reports are compiled, and stakeholders are informed about the incident and resolution measures to maintain trust and demonstrate a commitment to robust security practices (Figure 2.9).

2.5.1.3 Integration of real-time threat intelligence

Unfortunately, very little research has been conducted in this area (only a mere 1% of IS publications have discussed real-time intelligence at all) and the literature consists mostly of journalistic pieces, vendor whitepapers, or case studies, with just a few academic papers discussing the fusion between threat intelligence and risk management [59, 60].

The incorporation of continuous adaptation with real-time cyber risk management, as highlighted in the previous section, is significant in transcending the reactive and sluggish nature of traditional risk management methodologies toward proactive and agile resilience. However, for organizations to make calculated risk decisions based on an understanding of the likelihood and impact of an attack, RTTI must be integrated into this framework. At a basic level, RTTI involves the analysis and use of information about current and potential attacks to inform decisions and actions in the present. This will enhance the ability of organizations to understand their constantly changing risk posture, identify relevant threats and vulnerabilities, and allocate resources to best protect their assets [61].

- **Automated Threat Data Collection**: Utilize automated tools to gather and analyze threat data from multiple sources in real time.
- **Intelligence-Driven Decision-Making**: Employ AI and machine learning algorithms to process and interpret threat data, enabling timely and informed decision-making regarding threat responses.

Automated threat data collection will prevent severe bottlenecks during the intelligence-driven decision-making process by delivering more relevant and up-to-date information. Static intelligence gathers data on an as-needed basis and makes decisions using a fixed database, whereas dynamic intelligence collects a wide variety of information and stores it in a temporary data collection, making decisions by taking the current state of the data into account. By storing and making decisions against a real-time dataset, Global Data can provide the most accurate picture of current events, including the status and identities of all active hosts and connections to the network. High-quality threat data should include a wide variety of detailed information on security events from multiple sources so that informed decisions can be made to adequately respond to a threat. Automatic tools can gather a large amount of event data from firewalls, IDS/IPS, routers, and network applications without incurring the high overhead that would be involved with manual query and retrieval. Event data can later be correlated to form an accurate picture of security posture and aid in identifying changes in behavior by an attacker.

Threat data collection can take place at set intervals or continuously, with some options for an on-demand query to populate information such as new signature files as they become available or discovery of new devices on a network. Existing passive methods can be complex and require a level of expertise in network protocols and rule creation, for example, capturing NetFlow data and aggregating it and parsing syslogs to extract only relevant data. By employing an automated tool, we can capture a great deal of valuable context to establish a baseline of normal network behavior and to track changes in behavior that may identify an ongoing attack. Logs and alerts from multiple systems can be correlated to provide a higher level view of activity on a network and aid in identifying patterns of attack. Automatic analysis can help identify key events and produce meaningful reports with less manual effort. This is especially helpful in environments with a shortage of trained security staff.

2.5.1.4 Enhanced collaboration tools

Identification and access to specific expertise. A time-rich media such as blogs and wikis provide a useful platform for the sharing of risk knowledge, including specific threats and vulnerabilities and successful mitigation strategies. These tools can be used for the development of taxonomies and ontologies that further risk knowledge sharing. Expertise can be identified through the metadata associated with contributions to these media and targeted to help inform specific risk decisions or to assess the impact of a particular threat or vulnerability. Through the assessment of stakeholder risk assessments on an ongoing basis, the suitability and effectiveness of risk treatment options can be evaluated, helping guide the conduct of corrective actions and improving risk treatment decision-making.

Collaboration tools serve as an enabler for the sharing of risk-related data and knowledge across various stakeholders, including the technical, operational, and business communities. In doing so, these tools help to break down communication barriers that typically result from the use of different language and terminology and from organizational cultural differences. By connecting people to each other and to the data they need, collaboration tools facilitate the making of better informed and more timely risk management decisions. This can be realized through the following mechanisms:

- **Cross-Functional Teams:** Establish agile project teams that include members from IT, security, and operational departments to foster a holistic approach to risk management.

In today's increasingly interconnected and digitized world, effective CSRM has become a critical priority for organizations across various industries.

- **Communication Platforms:** Implement advanced communication tools to ensure seamless information sharing among stakeholders and rapid response capabilities.

These are two of the simplest factors in the framework, but they are essential to the success of a risk management strategy. Without the involvement of IT, security, and operations staff, it is impossible to have a comprehensive understanding of this diverse and complex system. Conversely, once cross-functional teams have been put in place, communication still tends to be segmented along departmental lines, which risks becoming silos. This can seriously hamper the ability to adapt to emerging system changes and threats, which may not be visible to all departmental perspectives. It is therefore a priority to enable transparent sharing of information in order to remain aware of all risk-related events that may affect the system. Rapid response capabilities represent the action component of an adaptation strategy, so a mechanism for translating knowledge into action must be put in place. An example would be the ability to automatically reallocate security resources to counter new threats, which may require changes to the operational configurations of the system. The idea is to refine these predictive capabilities, providing organizations with the tools to preemptively adjust their security measures according to the anticipated risk landscape, thus ensuring a proactive rather than reactive approach to cybersecurity [62].

2.5.1.5 Continuous improvement and learning

In support of both types of improvement activity, the framework will provide a set of risk management tools that can be used to determine how a change can be more effectively implemented, and a series of risk management research activities designed to increase knowledge about the elements of effective risk management and cyber risk specific.

Two basic types of improvement activity will be supported:

1. A series of informal strategic assessments designed to identify ways to improve the risk management approach.
2. More formal tests of change through the implementation of improved risk management processes and/or tools. These tests of change are experiments that are designed to increase knowledge about what works, what does not work, and how to apply specific risk management processes more effectively and/or tools. This knowledge will be directly fed back into the risk management processes and tools to ensure continuous improvement in their effectiveness.

The framework is built on the assumption that cybersecurity threat environments continue to evolve and thus requires corresponding development and evolution. Just as the threat environment is continuously changing, so too must the CSRM approach be continuously changing to remain effective. Continuous monitoring, continuous improvement, and continuous learning is the overarching theme for this part of the framework.

- **Feedback Loops:** Incorporate feedback mechanisms to learn from past security incidents and improve future responses.
- **Training and Awareness Programs:** Regular training sessions for all employees to keep them informed about the latest cybersecurity practices and threats.

Cybersecurity is often not at the forefront of employees' thoughts while going about their daily tasks. This can result in increased risk to the organization as employees may not be aware of the latest cybersecurity threats or have forgotten the details of risk treatment procedures for specific scenarios. Regular cybersecurity training and awareness sessions provide a means to keep cybersecurity in the minds of employees and will allow for the communication of changes to risk treatments or risk acceptance decisions. High-risk treatments may necessitate changes to the ways specific tasks are carried out, and employees need to be aware of these changes to ensure that there is no breach of risk treatment. A training program may also identify the point at which a risk treatment has become non-cost effective or identify new lower cost methods of reducing risk.

To ensure continuous improvement in cybersecurity response, it is essential to reflect upon past incidents and evaluate responses. A feedback loop system will allow for the collection of data from security incidents and near misses, which can then be fed back into the risk assessment and risk management phases. The feedback data will identify whether the implemented risk treatments were effective or not, and whether the incident response resulted in increased damage to the organization. Feedback data should be incorporated from all sources, and not just internal incidents. Publicized hacker methods and security breaches at other organizations provide a valuable source of learning, which should also be fed into the risk assessment and treatment phases. Real-time risk assessment will identify changes in the threat environment, which may result in the fast tracking of high-risk treatments.

2.5.1.6 Technology integration

But the most interesting is decision support system technology, which is very suitable with the idea of real-time monitoring in agile risk management. This technology provides an AI that can make logical decisions like a human. By storing any intelligence in a knowledge base, this system will produce better decisions, so there is a continuous learning process from the previous decision. This AI is very suitable with the continuous adaptation method by providing any result from the intelligence to the relevant place and time.

Related to the case of an agile method, which is a mix between incremental and iterative methods, there is a technology that can be used to write a scenario in each activity and store it in a model. So, the model will run automatically and provide a result based on the scenario. This technology can be used to support an agile method in assessing a risk. Simulation and modeling technology can also be used to simulate any action of the cyber-attack in the risk analysis phase. It is very suitable with an agile method because when we are using an iteration, it means we are repeating an action to achieve a different result. And the last is that the same action will have a different result in a different iteration compared to a cyber-attack action. So, we can identify any behavior of the cyber-attack by comparing the results from each iteration.

Technology and tools become the most important factors in CSRM. There are many existing technologies and tools that can be used to support CSRM activities. In general, the purpose of technology is to simplify the complexity of risk management activities and the achievement is to make risk management activities more efficient and effective.

CSRM is a different case compared to IT risk management in the past because cybersecurity is an activity to prevent and reduce the occurrence of cyber-attacks, which have a high complexity in terms of tools and spreading ways. To solve that problem, we must have technology that is able to predict and take preventive action against cyber-attack behavior.

Risk management and technology are two words that are deeply related. Technology is now one of the most advanced and fastest growing, which forces enterprises and organizations to engage in highly sophisticated and more complicated risk management activities. In general, this condition is affected by the idea that technology is a tool to achieve more profit. So, when organizations or enterprises do not want to be left behind by their competitors, they must implement the latest technology to support their business.

- **Advanced Analytics and Visualization Tools:** Use sophisticated analytics tools to visualize threat landscapes and model potential security breaches.

- **Cloud-based Security Solutions:** Leverage cloud technologies for scalable and flexible cybersecurity resources and infrastructure.

Rapid advancements in big data and machine learning areas have already been implemented into the cybersecurity domain. A good model must be able to capture known unknowns and unknown unknowns, as well as friendly and enemy probabilities. It should also be dynamic and enable various actions to change the game and mislead attackers. To achieve this, we need to be able to collect information about potential threats and vulnerabilities and input it into some form of model. Currently, the amount of information and variables involved in cybersecurity is too vast to visualize. Analyzing public news sources to understand the global threat landscape is a particularly difficult task. This is an area that would greatly benefit from big data analysis and visualization techniques. Big data analysis is a well-established field in computer science, but it has only reached the cybersecurity domain in recent years. A combination of big data analysis and visualization can provide a way to understand global threats. An example with too much data would be misleading, but data-driven models are the future of cybersecurity, and we must find ways to simplify and present this information. Big data analysis could reduce this down to several key points and model another variable, such as the probability of detection action. This would be a real tool for understanding and combating similar data exfiltration attacks.

2.6 DISCUSSION: EXPECTED OUTCOMES

The implementation of our new agile CSRM framework is poised to revolutionize conventional cybersecurity approaches by shifting from reactive to proactive risk management. By continuously assessing risks and aligning security investments accordingly, organizations can significantly reduce the frequency and impact of security incidents. This proactive stance not only prevents breaches but also facilitates a strategic perspective on risk acceptance, fostering new business opportunities and transforming the cybersecurity function from a cost center into a business enabler. Additionally, the framework enhances situational awareness at the senior management level, enabling well-informed decisions through a clearer comprehension of the risk landscape. This is achieved by integrating near RTTI and continuous risk management activities, which improve the organization's ability to swiftly adapt to and mitigate emerging threats. The framework's dynamic nature allows for the ongoing adjustment of security measures in response to evolving threats and vulnerabilities, ensuring optimal resource allocation and enhanced organizational resilience against cyber-attacks. Overall, this adaptive risk management model not only maintains security and safety levels but also promotes a systematic and prioritized approach to cybersecurity, making it a vital tool in today's fast-evolving digital world.

2.7 CONCLUSION AND OUTLOOK

The proposed agile CSRM framework, which integrates RTTI, is predicated on a comprehensive transformation of existing risk management practices. It facilitates organizations' development of the capacity to adapt to the dynamically evolving cyber threat landscape continuously and proactively. This framework, which merges cutting-edge technology, real-time intelligence, and strategic agility, transcends mere reactions to the limitations of traditional methods; it signifies a significant advance toward a more holistic and integrated comprehension of cybersecurity risks. By adopting this model, companies can reduce the likelihood of threats impacting their operations and reputation by preemptively identifying and adeptly responding to them. Future research should explore the integration of emerging technologies, such as AI and machine learning, to bolster the collection of real-time threat data and enhance decision-making processes. Moreover, investigations into the organizational and operational impacts of this framework will refine methodologies and ensure effective implementation, aligning with the strategic goals of organizations and safeguarding their most crucial assets in a rapidly evolving threat environment.

REFERENCES

[1] W. Ahmed, and M. Z. Rashidi, "Developing supply chain risk management capabilities by aligning strategies: Integrating Triple-A model," Measuring business excellence, 2022. researchgate.net

[2] M. Brand, V. Tiberius, P. M. Bican, and A. Brem, "Agility as an innovation driver: Towards an agile front end of innovation framework," Review of Managerial Science, 2021. [HTML]

[3] R. Shams, D. Vrontis, Z. Belyaeva, A. Ferraris, et al., "Strategic agility in international business: A conceptual framework for 'agile' multinationals," Journal of International Management, vol. 27, no. 1, Mar. 2021, Art. no. 100737, Elsevier. northumbria.ac.uk

[4] S. H. Björnsdóttir, P. Jensson, R. J. de Boer et al., "The importance of risk management: What is missing in ISO standards?" Risk Management, 2022, Wiley Online Library. researchgate.net

[5] F. R. Moreira, D. A. Da Silva Filho, and G. D. A. Nze, "Evaluating the performance of NIST's framework cybersecurity controls through a constructivist multicriteria methodology," IEEE Transactions on Engineering Management, 2021. ieee.org

[6] K. Stine, S. Quinn, G. Witte, et al., "Integrating cybersecurity and enterprise risk management (ERM)," National Institute of Standards and Technology, 2020 [Online]. Available: complexdiscovery.com. complexdiscovery.com

[7] F. Kitsios, E. Chatzidimitriou, and M. Kamariotou, "Developing a risk analysis strategy framework for impact assessment in information security management systems: A case study in IT consulting industry," Sustainability, 2022. mdpi.com
[8] I. Lee, "Cybersecurity: Risk management framework and investment cost analysis," Business Horizons, 2021. e-tarjome.com
[9] D. Landoll, The Security Risk Assessment Handbook: A Complete Guide for Performing Security Risk Assessments, CRC Press, Boca Raton, 2021. [HTML]
[10] C. Donalds, and K. M. Osei-Bryson, "Cybersecurity compliance behavior: Exploring the influences of individual decision style and other antecedents," Systems Journal of Information Management, vol. 51, Elsevier, 2020, 102056. [HTML]
[11] T. Muhammad, M. T. Munir, M. Z. Munir, et al., "Integrative cybersecurity: Merging zero trust, layered defense, and global standards for a resilient digital future," International Journal of Computer Science and Information Security, 2022 [Online]. Available: researchgate.net
[12] M. Antunes, M. Maximiano, R. Gomes, et al., "Information security and cybersecurity management: A case study with SMEs in Portugal," Journal of Cybersecurity and Privacy, 2021. mdpi.com
[13] B. Uchendu, C. Nurse, M. Bada, and S. Furnell, "Developing a cyber security culture: Current practices and future needs," Computers & Security, vol. 109, 2021, 102387. [PDF]
[14] F. A. Shaikh, and M. Siponen, "Information security risk assessments following cybersecurity breaches: The mediating role of top management attention to cybersecurity," Computers & Security, 2023. sciencedirect.com
[15] M. Eling, M. McShane, and T. Nguyen, "Cyber risk management: History and future research directions," Risk Management and Insurance Review, vol. 24, no. 3, 2021, 171–191. [HTML]
[16] A. A. Alahmari, and R. A. Duncan, "Investigating potential barriers to CSRMinvestment in SMEs," in 2021 13th International Conference on Electronics, Computers and Artificial Intelligence, 2021, pp. 1–6, IEEE, ieeexplore.ieee.org. researchgate.net.
[17] O. F. Keskin, K. M. Caramancion, I. Tatar, O. Raza et al., "Cyber third-party risk management: A comparison of non-intrusive risk scoring reports," Electronics, 2021. mdpi.com
[18] K. T. Kosmowski, E. Piesik, J. Piesik, and M. Śliwiński, "Integrated functional safety and cybersecurity evaluation in a framework for business continuity management," Energies, 2022. mdpi.com
[19] T. M. Popescu, A. M. Popescu, and G. Prostean, "IoT security risk management strategy reference model (IoTSRM2)," Future Internet, 2021. mdpi.com
[20] A. Cartwright, E. Cartwright, and E. S. Edun, "Cascading information on best practice: Cyber security risk management in UK micro and small businesses and the role of IT companies," Computers & Security, 2023. sciencedirect.com
[21] M. Kuhrmann, P. Tell, R. Hebig, J. Klünder, et al., "What makes agile software development agile?," IEEE Transactions on Software Engineering, vol. 47, no. 5, May 2021, 999–1002,. [PDF]
[22] D. Besson, "Principles of agile management by the manifesto and the Denning's developments–Remarks on what AM pretends to be," Przedsiębiorstwo zwinne w świetle badań empirycznych, p. 103, 2020, researchgate.net
[23] C. H. E. N. T. O. U. F. Zohair, "Software change decision support: A proposed framework and case study," Journal of the Chinese Institute of Engineers, vol. 39, no. 5, 2016, 531–537.
[24] P. Mall, R. Amin, A. K. Das, M. T. Leung, et al., "PUF-based authentication and key agreement protocols for IoT, WSNs, and smart grids: A comprehensive survey," IEEE Internet of Things Journal, vol. 9, no 11, 2022, 8205–8228. [HTML]

[25] C. J. Axon, and R. C. Darton, "Sustainability and risk–A review of energy security," Sustainable Production and Consumption, 2021. [HTML]
[26] M. I. Lunesu, R. Tonelli, L. Marchesi, and M. Marchesi, "Assessing the risk of software development in agile methodologies using simulation," IEEE Access, 2021. ieee.org
[27] P. Loft, Y. He, I. Yevseyeva, and I. Wagner, "CAESAR8: An agile enterprise architecture approach to managing information security risks," Computers & Security, 2022. sciencedirect.com
[28] K. Rindell, J. Ruohonen, J. Holvitie, S. Hyrynsalmi, et al., "Security in agile software development: A practitioner survey," Information and Software Technology, vol. 130, 2021, Elsevier. sciencedirect.com
[29] A. Naseer, H. Naseer, A. Ahmad, and S. B. Maynard, "Moving towards agile cybersecurity incident response: A case study exploring the enabling role of big data analytics-embedded dynamic capabilities," Computers & Security, vol. 135, Elsevier, 2023. sciencedirect.com.
[30] P. Radanliev, D. De Roure, K. Page, and J. R. C. Nurse, et al., "Cyber risk at the edge: Current and future trends on cyber risk analytics and artificial intelligence in the industrial internet of things and industry 4.0 supply," Cybersecurity, vol. 3. 2020, Springer. springer.com
[31] A. A. Mughal, "Building and securing the modern security operations center (SOC)," International Journal of Business Intelligence and Big Data, vol. 5, no 1, 2022, 1–15. research.tensorgate.org
[32] A. Yeboah-Ofori, S. Islam, S. W. Lee, et al., "Cyber threat predictive analytics for improving cyber supply chain security," in IEEE Access, vol. 9, pp. 94318–94337, 2021. ieee.org
[33] B. Shin, and P. B. Lowry, "A review and theoretical explanation of the 'Cyberthreat-Intelligence (CTI) capability' that needs to be fostered in information security practitioners and how this can be accomplished," Computers & Security, vol. 92, 2020, 101761. researchgate.net
[34] P. Gao, F. Shao, X. Liu, X. Xiao, Z. Qin, and F. Xu, "Enabling efficient cyber threat hunting with cyber threat intelligence," 2021 IEEE 37th International Conference on Data Engineering (ICDE), 2021. [PDF]
[35] H. Li, J. Wu, H. Xu, G. Li, and M. Guizani, "Explainable intelligence-driven defense mechanism against advanced persistent threats: A joint edge game and AI approach," IEEE Transactions on Dependable and Secure Computing, vol. 19, no 2, 2021, 757–775. [HTML]
[36] J. Burch, "Operationalizing intelligence collection in a complex world: Bridging the domestic & Foreign intelligence divide," Global Security & Intelligence Studies, 2022. scholasticahq.com
[37] A. Javed, A. Basit, F. Ejaz, A. Hameed, and Z. J. Fodor, "The role of advanced technologies and supply chain collaboration: During COVID-19 on sustainable supply chain performance," Discover Sustainability, vol. 5, no. 1, 2024, 46, Springer. springer.com
[38] A. Koohang, J. H. Nord, K. B. Ooi, G. W. H. Tan, et al., "Shaping the metaverse into reality: A holistic multidisciplinary understanding of opportunities, challenges, and avenues for future investigation," Journal of Computer Information Systems, vol. 63, Taylor & Francis, 2023. gre.ac.uk
[39] S. Saeed, S. A. Suayyid, M. S. Al-Ghamdi, H. Al-Muhaisen et al., "A systematic literature review on cyber threat intelligence for organizational cybersecurity resilience," Sensors, 2023. mdpi.com. mdpi.com
[40] C. Engel, S. Mencke, R. Heumüller, and R. Hormann, "Customizable operation center for smart security management," Procedia CIRP, vol. 96, 2021, 25–30. sciencedirect.com

[41] S. Samtani, M. Kantarcioglu, and H. Chen, "Trailblazing the artificial intelligence for cybersecurity discipline: A multi-disciplinary research roadmap," ACM Transactions on Management Information Systems, vol. 11, no 4, 2020, 1–19. acm.org
[42] N. Sun, M. Ding, J. Jiang, W. Xu, and X. Mo, "Cyber threat intelligence mining for proactive cybersecurity defense: A survey and new perspectives," IEEE Communications Surveys & Tutorials, 2023. ieee.org
[43] M. G. Kibria, N. I. Masuk, R. Safayet, H. Q. Nguyen, et al., "Plastic waste: Challenges and opportunities to mitigate pollution and effective management," Journal of Environmental Management, vol. 17, no 1, Springer, 2023, 20. springer.com
[44] M. Everard, P. Johnston, D. Santillo, and C. Staddon, "The role of ecosystems in mitigation and management of Covid-19 and other zoonoses," Environmental Science & Policy, vol. 111, 2020, 7–17, Elsevier. sciencedirect.com
[45] B. K. Singh, M. Delgado-Baquerizo, E. Egidi, et al., "Climate change impacts on plant pathogens, food security and paths forward," Nature Reviews, 2023 [Online]. Available: nature.com. nature.com
[46] N. Afzaliseresht, Y. Miao, S. Michalska, Q. Liu, et al., "From logs to stories: Human-centred data mining for cyber threat intelligence," in IEEE Access, vol. 8, pp. 19089–19099, 2020. ieee.org
[47] O. Kayode-Ajala, "Applications of Cyber Threat Intelligence (CTI) in financial institutions and challenges in its adoption," Applied Research in Artificial Intelligence and Cloud Computing, vol. 6, no 8, 2023, 1–21. researchberg.com
[48] W. Lam, A. Shi, R. Oei, S. Zhang, and M. D. Ernst, "Dependent-test-aware regression testing techniques," in Proc. of the International Symposium on Software Testing and Analysis, 2020, pp. 1–11. dl.acm.org. acm.org
[49] P. Radoglou-Grammatikis, K. Rompolos, et al., "Modeling, detecting, and mitigating threats against industrial healthcare systems: A combined software defined networking and reinforcement learning approach," IEEE Transactions on Industrial Informatics, 2021. ieee.org
[50] L. Schneller, C. N. Porter, and A. Wakefield, "Implementing converged security risk management: Drivers, barriers, and facilitators," Security Journal, 2023. springer.com
[51] R. Hoffmann, J. Napiórkowski, T. Protasowicki et al., "Risk based approach in scope of cybersecurity threats and requirements," Procedia Computer Science, vol. 176, 2020, 2810–2819. sciencedirect.com
[52] S. Jarjoui, and R. Murimi, "A framework for enterprise cybersecurity risk management," Advances in cybersecurity management, 2021. researchgate.net
[53] S. H. Alsamhi, A. V. Shvetsov, S. Kumar, S. V. Shvetsova, et al., "UAV computing-assisted search and rescue mission framework for disaster and harsh environment mitigation," Drones, 2022 [Online]. Available: mdpi.com mdpi.com
[54] M. S. A. Lee, L. Floridi, and A. Denev, "Innovating with confidence: Embedding AI governance and fairness in a financial services risk management framework," in Ethics, Governance, and Policies in Artificial Intelligence, Springer, 2021. researchgate.net
[55] C. H. Koo, S. Schröck, M. Vorderer, J. Richter et al., "A model-based and software-assisted safety assessment concept for reconfigurable PnP-systems," Procedia CIRP, 2020. sciencedirect.com
[56] M. Pfannemüller, M. Breitbach, M. Weckesser et al., "REACT-ION: A model-based runtime environment for situation-aware adaptations," in Proc. of the International Conference on Autonomic and Adaptive Systems, 2021, pp. 1–6. acm.org
[57] H. Salin, and M. Lundgren, "Towards agile CSRM for autonomous software engineering teams," Journal of Cybersecurity and Privacy, 2022. mdpi.com

[58] A. Mathrani, S. Wickramasinghe, and N. P. Jayamaha, "An evaluation of documentation requirements for ISO 9001 compliance in scrum projects," The TQM Journal, 2022. researchgate.net
[59] E. B. Hansen, and S. Bøgh, "Artificial intelligence and internet of things in small and medium-sized enterprises: A survey," Journal of Manufacturing Systems, 2021. researchgate.net
[60] J. Laguarta, F. Hueto, and B. Subirana, "COVID-19 artificial intelligence diagnosis using only cough recordings," IEEE Open Journal of Engineering in Medicine and Biology, 2020. ieee.org
[61] D. Schlette, M. Caselli, and G. Pernul, "A comparative study on cyber threat intelligence: The security incident response perspective," IEEE Communications Surveys & Tutorials, 2021. uni-regensburg.de
[62] S. Kumar, and R. Anbanandam, "Impact of risk management culture on supply chain resilience: An empirical study from Indian manufacturing industry," Proceedings of the Institution of Mechanical Engineers Part O Journal of Risk and Reliability, 2020, journals.sagepub.com. researchgate.net

Chapter 3

Advancing cybersecurity

Machine learning algorithms for intrusion detection systems

Nessrine Moumen

3.1 INTRODUCTION

In practice, it becomes increasingly difficult to find impenetrable computer systems. Indeed, attacks on these platforms are constantly increasing in number and intensity. At the same time, interconnected networks are experiencing a rising increase [1]. This phenomenon is further accentuated with the advent of the Internet of Things (IoT), which aims to bring together all kinds of equipment with a wireless interface [2]. However, these emerging networks rarely incorporate effective protections against attacks by cybercriminals. These pose a real threat to both users and businesses. It is therefore necessary to find solutions capable of effectively securing these objects while respecting their hardware and software constraints [3].

One of the first steps to improve the security of computer systems is to be able to detect attack attempts as soon as they occur [4]. For this, it is necessary to look at past attacks involving the host or network but also to prevent new types of attacks.

Within this context, it becomes more important to be capable of detecting and preventing different kinds of attacks. To this end, intrusion detection systems (IDSs) are used. An IDS can alarm administrators of malicious behavior [5]. Most IDSs require an amount of manual maintenance to get decent results. The main concern of this chapter is to figure out if an IDS with satisfactory performance can operate. This is achieved by using machine learning algorithms. These algorithms can learn from input data and find models that would seem to be well applicable to the question of self-learning intrusion detection [6]. Therefore, machine learning algorithms seem promising in terms of security for the automated detection of intrusion issues [7].

The reminder of this chapter is organized as follows. Section 3.2 examines what an IDS is. Section 3.3 presents a set of metrics commonly used to evaluate the performance of IDSs. Section 3.4 lists some relevant machine learning classifiers (MLCs). In Section 3.5, I describe how I implemented the chosen machine learning model. This section encompasses and discusses the obtained. Finally, Section 3.6 concludes the chapter.

 DOI: 10.1201/9781003478676-3

3.2 INTRUSION DETECTION SYSTEM

IDS stands for intrusion detection system. It refers to systems used to monitor a given network or host's activity to detect and possibly react to any intrusion attempt [1, 8]. Intrusion detection is the process of monitoring and analyzing intrusion events that occur on a computer or on a network. These kinds of events are defined as attempts that compromise confidentiality, integrity or availability or exceed a computer or network's security conditions.

3.2.1 Classification of IDS

Nowadays, a variety of available IDS are endowed with intelligent tools to monitor and analyze systems. As illustrated in Figure 3.1, each approach has its own distinct advantages and drawbacks. All approaches can be interpreted in terms of an IDS model, including [9]

- The location of IDS
- Detection methods
- Types of response
- Frequency of use

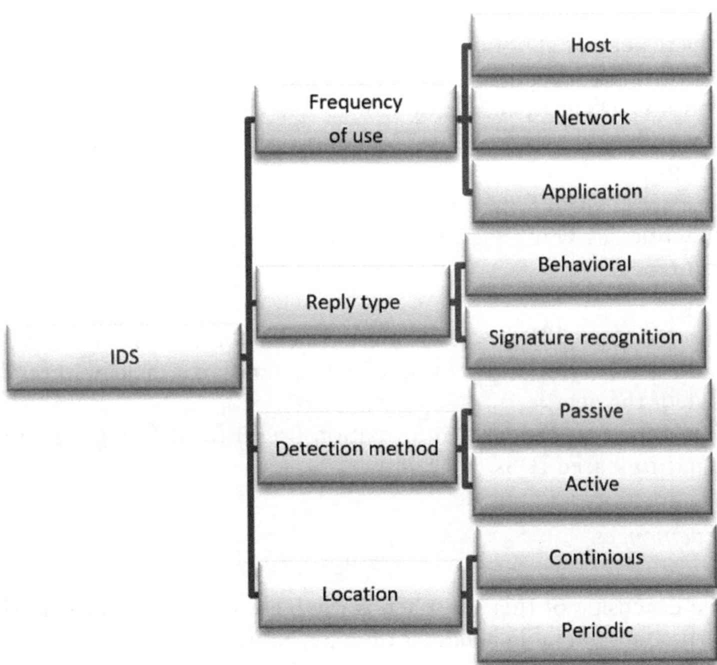

Figure 3.1 Classification of IDS. (Created by the author.)

3.2.2 Detection modes

Practically, intrusion detection mechanisms can be categorized into two distinct modes: anomaly detection and signature-based detection. The latter is more commonly employed as the principal operational mode in IDSs. However, contemporary advancements in this field are seeing a trend toward the integration of both modes to enhance the precision of intrusion detection.

3.2.2.1 Anomaly detection

It consists in detecting anomalies in relation with "usual traffic" profile. The implementation always includes a learning phase during which the IDS will "discover" the monitoring elements' "normal" operation [10]. They are thus able to report discrepancies from the reference operation.

In the case of HIDS (host-based IDS), this type of detection can be based on information such as the rate of CPU use, disk activity, connection times, or the use of certain files (hours at the office, etc.) [11]. As an example, a project was sponsored by DARPA IDS (in 1998 and 1999), in collaboration with the Lincoln laboratory at MIT [12]. This project aims to provide a meaningful set of learning data, including background traffic and intrusive activities (i.e., intrusive traffic or system events caused by attacks). Background traffic was derived from statistical data collected on the Air Force base network, while attacks were generated not only by specially created scripts but also by scripts collected through specialized sites and mailing lists. The data collected concerned both HIDS (e.g., audit data from Solaris stations, disk dump of UNIX machines and BSM "Basic Security Monitoring") and NIDS (network-based IDS) [13, 14].

The test set DARPA'98 used around 300 attacks (classified into 38 types of attacks), while DARPA'99 used around 50 types of attacks [15, 16].

Advantages

- Anomaly-based IDSs detect unusual behavior and thus have the capacity to detect symptoms of known and unknown attacks without knowledge of the details.
- This approach produces information useful for defining signatures for signature-based IDSs.

Inconveniences

- The downside of this approach is the large number of false alarms due to the unpredictable behavior of network users.
- It often requires a long-term history of recorded events to characterize normal patterns of behavior. Systems based on this approach must have some intelligence for machine learning.

3.2.2.2 Signature recognition

This approach consists of searching the activity of the monitored system for the fingerprints (or signatures) of known attacks [17]. This type of IDS is purely reactive; it can only detect attacks of which it has the signature before. Therefore, it requires frequent updates.

In addition, the effectiveness of this detection system strongly depends on the precision of its signature base. This is why these systems are bypassed by hackers who use so-called escape techniques, which consist of making up the attacks used. These techniques tend to vary the signatures of attacks, which are therefore no longer recognized by the IDS.

It is possible to develop more generic signatures, which make it possible to detect variants of the same attack, but this requires a good knowledge of the attacks and of the network, to stop the variants of an attack and not to interfere with normal network traffic. A signature is used to define the characteristics of an attack, at the packet level (up to Transmission Control Protocol (TCP) or User Datagram Protocol (UDP)) or at the protocol level (Hypertext Transfer Protocol (HTTP), File Transfer Protocol (FTP), etc.).

At the packet level, the IDS will analyze the different parameters of all packets in transit and compare them with the signatures of known attacks. At the protocol level, the IDS would check if the commands sent are correct or do not contain an attack. This functionality has mainly been developed for HTTP.

This example seemed very instructive since it shows the role of signatures and their creation. Be aware that IDS vendor sites offer signature updates based on new identified attacks.

However, the more different signatures to test, the longer the processing time, so the use of more sophisticated signatures can save significant time.

It should be noted that, a poorly crafted signature can ignore actual attacks or identify normal traffic as an attack. It is therefore advisable to handle the creation of signatures with care and having good knowledge of the monitored network and existing attacks. This IDS is very effective in detecting attacks without producing many false alarms. Moreover, it can quickly and surely diagnose the use of a specific tool or attack technique [18].

Accordingly, this can help security officials prioritize corrective action. However, it can only detect known attacks, the signatures of which are entered into the system; therefore, the detection system must be constantly updated with the signatures of new attacks. It is also noticed that many systems adopting this approach are designed to use a limited number of signatures that can be defined, which prevent them from detecting variants of these attacks.

3.3 EVALUATION CRITERIA

Even though IDSs have become the ubiquitous defense tools in today's computer systems and machines, so far, we do not have a rigorous scientific methodology for examining the efficiency and effectiveness of these systems. In this section, a set of partial measures that can be used to test the performance of IDS is given.

3.3.1 Classification errors

IDS generates a variety of errors that have an impact on its performance [7]. Figure 3.2 highlights these scenarios; the true positives occur when security policies are violated. The true negatives are instances where no alarm sounds and nothing unusual occurs [19]. False positives (FPs) happen when an alarm is triggered even though nothing abnormal has occurred. False negatives are cases that occur when an alarm fails to go off when something unusual occurs. At first appearance, one might believe that a FP is less harmful than a false negative [20].

The terms positive and negative in a binary task apply to the conformity of the prediction to the external assessment (sometimes called observation). With these meanings, the following matrix is constructed, shown in Table 3.1 and called the "confusion matrix." It serves as a tool used to evaluate the performance of a classification model and assumes a tabular format, offering a visual representation conducive to analytical examination. This matrix facilitates the comparison between the actual target values and those predicted by the machine learning model, also called a "classifier."

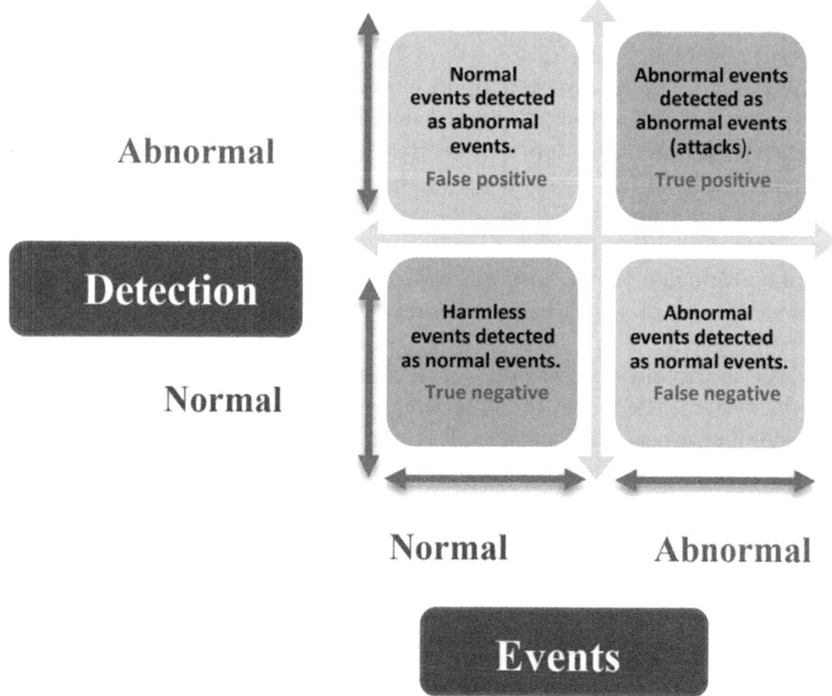

Figure 3.2 Types of IDS's detections. (Created by the author.)

Table 3.1 Evaluation of binary classifiers (created by the author)

	Actual class (observation)	
Predicted class (expectation)	TP (true positive) Correct result	FP (false positive) Unexpected result
	FN (false negative) Missing result	TN (true negative) Correct absence of result

In this context, recall, precision, accuracy, and F-measure can all be defined:

$$\text{Recall} = \frac{TP}{TP + FN} \quad (1)$$

$$\text{Precision} = \frac{TP}{TP + FP} \quad (2)$$

$$\text{Accuracy} = \frac{TP + TN}{TP + FP + TN + FN} \quad (3)$$

F-measure (also known as the F1-score or F-score) is an accuracy test metric that is defined as the weighted harmonic mean of the test's precision and recall:

$$\text{F} - \text{measure} = 2\ .\frac{Precision\ .Recall}{Precision + Recall} = \frac{TP}{TP + \frac{1}{2} + (FP + FN)} \quad (4)$$

Otherwise, the ratio of the total number of correctly classified samples divided by the total number of samples is known as accuracy. The confusion matrix is a graphical representation of a machine learning model's efficiency. In a classification problem, it summarizes the performance of each class.

3.4 MACHINE LEARNING CLASSIFIERS' OVERVIEW

Machine learning encompasses the development of computational models derived from empirical data using algorithms [21]. These models aim to generalize by representing or approximating the data, enabling the prediction of unknown values, and enhancing understanding of existing ones. Machine learning finds applications in various fields, such as finance, cybersecurity (for intrusion detection), personalized search engines, antivirus software, and cryptanalysis. The typical lifecycle of a machine learning implementation consists of the following steps, as illustrated in Figure 3.3:

1. Data acquisition
2. Data pre-processing
3. Training
4. Validation
5. Deployment.

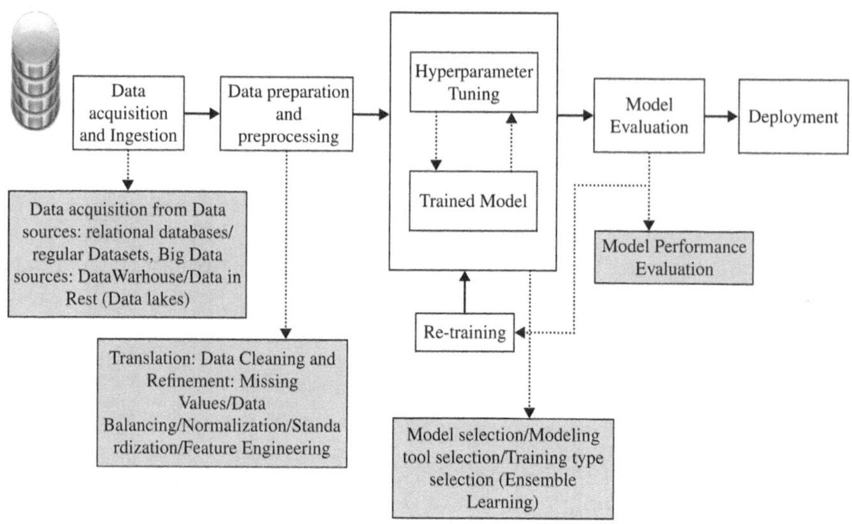

Figure 3.3 Machine learning pipelines: Streamlining the model development process. (Created by the author.)

This section provides an introduction of some machine learning techniques and demonstrates the need for implementing machine learning models in various areas, including intrusion detection [22]. The continual advancement of technology highlights the rising need to implement machine learning algorithms for the analysis and extraction of insights from massive datasets. In general, machine learning algorithms can be classified as supervised and unsupervised algorithms [23]. A supervised algorithm learns from pre-labeled (classified) objects to predict the object class. On the other hand, the unsupervised algorithm identifies essential clusters from unlabeled data, facilitating the natural grouping of objects.

Various types of models can be distinguished based on their supervised or unsupervised nature, as well as their classification or regression functionality. Classification models categorize data, while regression models predict values associated with each data point [24].

Since the IDS challenge is related to a classification problem, this study focuses only on investigating supervised/classification machine learning algorithms because, to the best of my knowledge, supervised deep learning algorithms find less application in intrusion detection, especially since spam detection doesn't rely on supervised deep learning algorithms. Despite its relationship to natural language processing (NLP), there is no supervised deep learning algorithm applied to spam detection. Hence, this gap opens many research opportunities.

The dataset employed in this research comprises predefined classes. Presented below is a non-comprehensive selection of diverse classifiers that may be contemplated for the implementation of an intelligent IDS.

3.4.1 Naïve Bayes

This model employs Bayes' theorem as its foundational framework. A fundamental assumption underlying this method is the independence of each class of examples. During the learning phase, the algorithm computes the probabilities associated with each class as well as the probabilities of each attribute given a particular class. This computation entails examining the frequency of occurrence of each class and the values of attributes within each class within the training dataset. Subsequently, during the execution phase, the algorithm uses these calculated probabilities to determine the most likely class assignment for an unknown instance based on its attributes [25]. A particular advantage of this method is its efficacy in requiring only a limited amount of information during the learning phase [26]. Mathematically, it is based on Bayes' theorem:

$$P(B|A) = P(B \cap A) \div P(A)$$

$$= [P(A|B) * P(B)] \div P(A) \tag{5}$$

One key condition that must be fulfilled by the undermentioned equation is that neither (B) nor (A) should equal zero. Therefore, this condition must be fulfilled as the Bayes theorem is extended in all situations. In addition, the alternate version of the Bayes theorem is usually used where two opposing arguments or theories are examined:

$$P(B|A) = P(A|A) * P(B)/P(A|B) * P(AB) + P(A|B') * P(B) \tag{6}$$

The related chance of the original degree of belief against A is P(A), where P(A') = 1 – P(A). The extended form of the Bayes Theorem for every partition {Ai} of the sample space is

$$P(B|A) = P(A|B) * P(B)/\Sigma I P(Ai|B) * P(B) \tag{7}$$

3.4.2 K-nearest neighbors

K-nearest neighbors (KNN) model belongs to the instance-based learning family. It seeks the KNN of the example to be predicted, subsequently returning the majority class among these neighbors as the response for this example [27]. Distance calculation can be performed using Euclidean or alternative distance metrics. Learning is straightforward as it entails memorizing the examples. Despite this, prediction may incur substantial computational overhead due to the necessity of identifying the KNN in a potentially high-dimensional space.

3.4.3 Learning decision trees

Decision trees create classification or regression models represented as tree structures, breaking down datasets into smaller subsets while incrementally

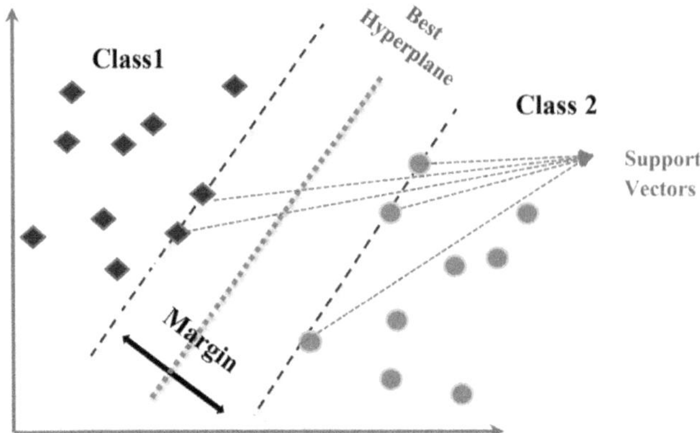

Figure 3.4 Verification of the margin between hyperplanes and data positions. (Adapted by the author from [30].)

constructing an associated decision tree. The resulting tree consists of decision nodes and leaf nodes, where decision nodes have two or more branches, and leaf nodes denote classifications or decisions [28]. The top decision node, coinciding with the best predictor, is called the root node. Decision trees can handle both categorical and numerical data, offering versatility in predictive modeling and ease of interpretation.

3.4.4 Support vector machine

Support vector machine (SVM) is a supervised model of machine learning, which identifies a hyperplane among the data represented, as visually depicted in Figure 3.4. Data can be represented within multidimensional space. Thereby, SVMs are widely used in models of classification [29]. An SVM will use the hyperplane that best separates the different classes. In some case scenarios, when we have different hyperplanes separating different classes, the proper identification of the hyperplane is determined using what is termed a margin or space. The margin is the closest distance between the hyperplanes and the data positions. You can consult the following representation to check the margin.

3.5 RELATED LITERATURE

An extensive amount of literature has been produced on NIDS [9, 24] and HIDS. These research studies are inspired by challenges related to managing tremendous amounts of local and network data, the shifting character of data, and limited features and detection capabilities of existing antiviruses [31].

In the literature, machine learning-based IDS algorithms and techniques frequently generate results with relatively limited datasets aimed at boosting network security. This provides no clear indication of the attack detection capacity of IDS datasets.

Many literature surveys gather or integrate the current state of the art. However, while several of these studies provide a detailed analysis, they do not include a specific practical case study or application of a cyber detection problem, such as spam detection. Beyond that, certain papers lack a specific reference to machine learning [33], while others do not address cybersecurity.

This critical analysis makes the proposed machine learning-based IDS an innovative and comprehensive alternative for efficiently tackling intrusion detection issues.

3.6 IMPLEMENTATION, RESULTS, AND DISCUSSION

In this section, the implementation will be a SPAM filter with a Naive Bayes classifier and Python. It is a spam detector. Spam refers to the use of digital communications services to transmit messages that are not requested or unwanted.

Our proposed MLC is shown in Figure 3.5. The main constructs of this framework are as follows:

The underlying principle of this classifier is simple: to identify words commonly associated with spam. The aim is to construct a binary classifier capable of discerning between spam and non-spam emails using Python in conjunction with the Natural Language Toolkit (NLTK) library. NLTK, a Python package, facilitates the handling and manipulation of extensive textual data for developers and data scientists alike. The Bayesian theorem is extensively employed in classification tasks, including spam filtering. In this study, a Bayesian classifier is used to identify spam messages.

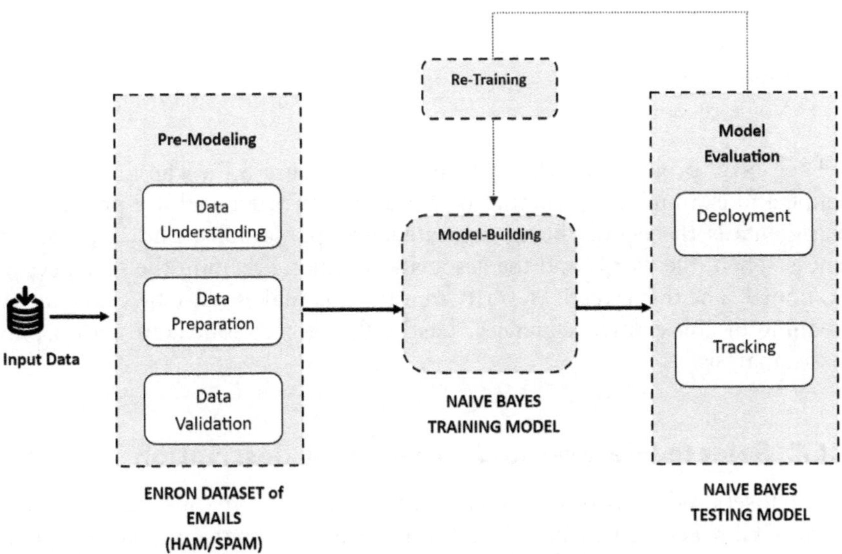

Figure 3.5 Steps involved in MLC for SPAM detection. (Created by the author.)

Particularly, the occurrence of spam within a message is defined as the event of interest, signaled by the presence of certain flagged words (e.g., "Will not believe your eyes," "you have won," or "This isn't spam"). The formulation, as extracted from Wikipedia, considers the probability of a message being spam given the presence of these flagged words:

$$P(spam|words) = P(words|spam).\ P(spam)\ P(words). \tag{6}$$

Advantages:

- It is very quick and efficient to use.
- Need less data from preparation.
- Allows assumptions of probabilism.
- Covers continuous and discrete data.
- It is a generative model, i.e., even if some feature is unavailable by altering decision rules, it will make predictions.

3.6.1 NLP—Natural language processing

NLP is a field of computational linguistics that aims to enable machines to understand, manipulate, and generate natural language. NLP is positioned at the intersection of linguistics and computer science, facilitating direct interaction between computers and humans. It empowers machines to analyze, interpret, and identify meaning from natural language, with application. NLP is frequently used in many areas and applications, such as:

- Real-time translation
- Instant automatic synthesis
- Opinion mining
- Voice recognition

Each NLP project goes through five steps (Figure 3.6). The first step is to define and examine the structure of words to construct an NLP project. This phase entails the details being separated into paragraphs, phrases, and sentences. Then, the words and the associations among them in the sentences are examined. The third step is to verify that the text makes sense by analyzing the meaning of consecutive sentences. Lastly, the project concludes with a pragmatic analysis.

3.6.2 Selected dataset and experiment description

The dataset used for this research paper was chosen from the dataset provided by the UCI Machine Learning Repository. Available on the website at: https://www2.aueb.gr/users/ion/data/enron-spam/index.html.

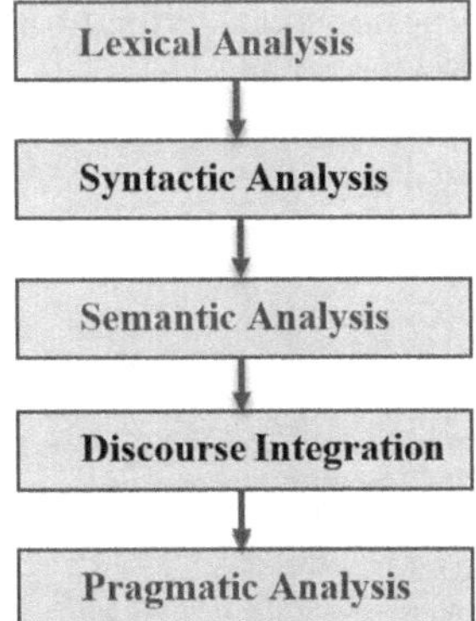

Figure 3.6 Verification of the margin between hyperplanes and data positions. (Created by the author.)

Basically, the website provides four datasets:

- Ling-spam
- PU1
- PU123A
- Enron-spam

This project uses the Enron spam dataset. The training dataset contains 1700 normal emails and 400 abnormal emails.

Naive Bayes classifier's first step is to determine the total number of classes (outputs) and calculate the conditional probability for each class of the datasets. Afterward, the conditional probability for each attribute would be calculated. In the previous section, the standard formula of Naive Bayes was provided.

After preparing the dataset by randomly shuffling email lists, the first step is text pre-processing. To that end, two functions are imported from the NLTK library, which is a Python package that helps to manage and manipulate large amounts of text for developers and data scientists.

The first is the tokenization. In natural language processing terminologies, tokenization essentially consists of breaking data down into smaller pieces or tokens, so that they can be easily analyzed.

The second function is lemmatizer; the objective of lemmatization is to connect words that have different forms because it considers the morphological analysis of words. The result of lemmatization is an appropriate word. Pre-processing is the most impactful step toward building our machine learning model for SPAM detection. Pre-processing plays an important role to achieve classification accuracy, which improves the effectiveness and efficiency of the IDS.

To process the emails, functions are defined for performing lemmatization and tokenization on our dataset:

```
wn = nltk.WordNetLemmatizer()
import re

def lemmatizer(text):
    text = [wn.lemmatize(word) for word in text]
    return text

def tokenization(text):  # Tokenization function
    text = re.split('\W+', text)
    return text
```

The second step is the extraction of the features using tf-idf information retrieval technique:

```
from sklearn.feature_extraction.text import CountVectorizer,
TfidfVectorizer, TfidfTransformer

count_vectorizer = CountVectorizer(analyzer="word",tokenizer=
nltk.word_tokenize, stop_words='english', max_df= 0.5,
min_df=2,ngram_range=(1,3), max_features=None)

bag_of_words = count_vectorizer.fit_transform(emails)
email_vectorized= count_vectorizer.transform(emails)
emails = pd.DataFrame(email_vectorized.toarray(),
columns=count_vectorizer.get_feature_names())
```

Definition of the training model's Python function:

```
from sklearn.model_selection import train_test_split
from sklearn.naive_bayes import GaussianNB
from sklearn.metrics import accuracy_score

train_x, test_x, train_y, test_y = train_test_split(emails,
labels, test_size = 0.3)

classifier = GaussianNB()  # Import Model
```

In this paperwork, to detect an anomaly and to achieve the highest detection accuracy with the shortest possible time, the Naive Bayes classifier was performed as the classification algorithm to find the detection accuracy and the FP:

```
classifier.fit (train_x, train_y) # Launch Training
```

The conditional probability theorem will be applied to calculate the probability that a message is SPAM based on the frequency of occurrence of words in the email.

Finally, for performing the classification, the highest probability for each email is compared.

3.6.3 Results and discussions

After the creation of the training models, the next step is the testing phase process and evaluation. Here is the Python code snippet for model testing/evaluation:

```
predictions = classifier.predict(test_x) # Predicting on Test data
# Calculate accuracy
print("accuracy of the model is {}".format(classifier.score
(test_x, test_y)))
# Calculate precision
print("precision score of Naive Bayes is{}".format(metrics.
precision_score(test_y, predictions, average="weighted")))
```

A machine learning model's performance can be assessed using evaluation metrics. The following performance metrics are computed:

- FP: This value represents an incorrect classification decision where the normal email (Ham) classified as Spam email.
- Detection accuracy (precision): It is one of the primary performance indicators. It is the ratio of the number of samples correctly classified, divided by the total number of samples.

Machine learning-based IDS aim to improve upon the adaptability and accuracy of existing systems, especially in terms of dealing with the dynamic and high-volume data characteristic of network environments by initially training on historical data to learn normal and malicious patterns. Once trained:

Step 1—data collection and pre-processing: The system collects network traffic in real-time, which cludes packets, flow data, or system logs. It preprocesses the data to extract relevant features while potentially also filtering out noise and irrelevant information.
Step 2—model prediction: The trained model then applies to this preprocessed data in real-time, classifying the traffic into either normal or suspicious (anomalous) behaviors.
Step 3—anomaly detection: When an anomaly is detected—which could be a signature of an intrusion or an attack—the system flags it for further investigation.
Step 4—adaptive learning: Machine learning-based IDS is also smart; it can learn from new patterns in real-time data to stay updated with current threat landscapes, adjusting its detection algorithms accordingly.
Step 5—real-time response: Upon detection of a potential threat, the system immediately responds—sending alerts to administrators, blocking suspicious traffic, or adapting firewall rules.

```
Training accuracy : 98.36 %
Test accuracy : 97.54 %

Test set confusion matrix:
Accuracy : 0.9754098360655737
F1 score : 0.9703572334417785
Recall   : 0.9489593667086224

classifocation report :

                precision   recall   f1-score    support

             0.96         0.99      0.98          2740
             0.99         0.95

acuracy                            0.98           4758
macro avg    0.98       0.97       0.97           4758
wighted avg  0.98       0.98       0.98           4758


confusion matrix :
  [[ 2726     14]
  [  103     1915]]
```

Figure 3.7 Naive Bayes evaluation. (Source: Author.)

In this section, the implementation of the proposed algorithm is carried out to validate its performance (Figure 3.7). The experiment has conducted that the proposed machine learning algorithm is able to differentiate between normal and malicious email. Two metrics have been evaluated for the proposed method: training accuracy was 98.36%, and the test accuracy for naïve Bayesian classifiers was 97.54%, which is high classification accuracy, with a 12.21% for false positive percentage. It clearly proves that our model has provided a high accuracy rate.

```
# Output Evaluation Metrics
From sklearn.metrics  import precision_score,recall_score,
confusion_matrix, classifition_report, accuracy_score, f1_score

print('Accuracy:', accuracy_score(y_test, predictions_test))
print('F1 score:', f1_score(y_test, predictions_test))
print('Recall:', recall_score(y_test, predictions_test))
print('Precision:', precision_score(y_test, predictions_test))
print('\n classification report:\n', classification_report
(y_test,predictions_test))
print('\n confusion matrix:\n',confusion_matrix(y_test,
predictions_test))
```

A concrete analysis shows that for every 100 emails classified as potentially malicious by the IDS, approximately 12 are FPs, based on the obtained FP rate of 12.21%. This indicates that while the IDS is highly accurate in distinguishing between normal and malicious emails, a noticeable proportion of normal emails are misclassified as spam. The potential causes of these FPs include ambiguous email content, variations in language or writing style, and noise in the dataset leading to misinterpretation by the classifier. To mitigate this percentage, several strategies can be implemented. First, refining feature selection criteria and optimizing IDS algorithms to better capture the distinguishing characteristics of malicious emails can help reduce FPs. In addition, implementing post-processing filters or validation mechanisms to verify suspicious detections before raising alerts can improve IDS accuracy while minimizing FPs. Furthermore, the use of feedback mechanisms from users or security analysts to provide corrective input on misclassified emails can facilitate continuous learning and refinement of the IDS over time, ultimately improving its performance in real-world scenarios.

3.6.4 Classification accuracy and precision of three other classifiers

The following observations are made from three other conventional classifiers. It is finally observed that the Naive Bayes classifier produces better performance than others.

In this implementation section, some other popular classification techniques have been used for a comparative study and are briefly presented as follows:

```
from sklearn.neighbors import KNeighborsClassifier
from sklearn.svm import SVC
from sklearn.ensemble import RandomForestClassifier

KNN evaluation: Test accuracy: 96.40 %,
                Precision: 94.57 %

SVM evaluation: Test accuracy: 95.99 %,
                Precision: 97.00%

Random Forest evaluation: Test accuracy: 84.88 %,
                          Precision: 89.06 %
```

3.7 CONCLUSION

Data is a sensitive and critical resource that requires special processing, especially when it is shared. In this context, security mechanisms emerge as an essential asset to facilitate data sharing while maintaining the confidentiality of the data transmitted. Malware poses a widespread and persistent cyber threat, presenting ongoing security challenges for modern organizations. Hackers are constantly improving, so classic techniques for detecting intrusion are outdated. Therefore, machine learning techniques can help with malware detection.

This study identifies a variety of classification algorithms, which are considered one of the basic data mining functions since they correctly anticipate the target class in the data for each event. The proposed solution was a machine learning algorithm for intrusion detection using a naive Bayesian classifier, which performs spam detections. The proposed algorithm solves certain difficulties in data mining, such as continuous attribute management and noise reduction in the training dataset. The experiment has demonstrated that it offers very good detection performance with experimental validation.

The classification decision was made within an acceptable timeframe for Naive Bayes since it performs a learning phase and generates a model, which it then uses to make the classification; the accuracy of the approach is very satisfactory in terms of the detection rate. Despite this, the parameter of FP is relatively high.

Furthermore, to improve the proposed solution, it is necessary to consider the other performance evaluation parameters, the average accuracy rates are not sufficient to detect the intrusion. False negative and false positive rates should also be taken into consideration.

In the context of the scalability of machine learning-based IDS, it is essential to consider the ability of the system to handle the large and growing volume of network data that is common in larger and more complex environments. Scalability is essential to ensure that detection capabilities can grow with the size of the network and the diversity of threats. Machine learning classifiers must be designed to maintain high performance without a proportional increase in computational cost or resource requirements. However, the evolution of such systems often leads to significant processing overhead, particularly when dealing with high-dimensional feature spaces and real-time data analysis. Optimizations, such as feature selection, dimensionality reduction, and efficient algorithm design, can mitigate the computational cost. Additionally, distributed computing architectures can be used to process data in parallel, improving the throughput and scalability of the system. Furthermore, proactive monitoring and dynamic resource allocation strategies enable the IDS to adapt to evolving workloads and allocate resources effectively, ensuring adaptive operation in diverse network environments. Despite these techniques, the goal remains challenging: achieving an effective balance between accuracy, speed, and resource consumption, which is critical for deploying machine learning classifiers in advanced IDS over extensive network infrastructures [34].

The future work or next phase of this project could focus on addressing the challenges identified and further enhancing the proposed machine learning-based IDS. One aspect of future work that could involve exploring the integration of advanced predictive analytics algorithms and threat intelligence feeds into the IDS framework could enhance its predictive capabilities and enable proactive defense mechanisms. This potential transition from intrusion detection to intrusion prediction systems (IPS) aims to go beyond intrusion detection approach by not only identifying ongoing security threats but also predicting and preemptively mitigating potential future intrusions before they occur. It is important to note that research into the development of adaptive IPS systems that constantly learn from new input and adjust their predictive models in real time could contribute to more effective intrusion prevention and response systems. To sum up, the potential move toward IPS represents a future direction in the field of cybersecurity, offering the prospect of proactive defense mechanisms against evolving cyber-attacks.

REFERENCES

1. Ahmad, Z., Shahid Khan, A., Wai Shiang, C., Abdullah, J., Ahmad, F.: Network Intrusion Detection System: A Systematic Study of Machine Learning and Deep Learning Approaches. Trans. Emerg. Telecommun. Technol. 32 (2021). https://doi.org/10.1002/ett.4150
2. Mehmood, Y., Ahmad, F., Yaqoob, I., Adnane, A., Imran, M., Guizani, S.: Internet-of-Things-Based Smart Cities: Recent Advances and Challenges. IEEE Commun. Mag. 55, 16–24 (2017). https://doi.org/10.1109/MCOM.2017.1600514
3. Roman, R., Zhou, J., Lopez, J.: Applying intrusion detection systems to wireless sensor networks. In: 2006 3rd IEEE Consumer Communications and Networking Conference, CCNC 2006, pp. 640–644 (2006).
4. Tarter, A.: Importance of Cyber Security. In: Bayerl, P.S., Karlović, R., Akhgar, B., and Markarian, G. (eds.) Community Policing – A European Perspective: Strategies, Best Practices and Guidelines, pp. 213–230. Springer International Publishing, Cham (2017).
5. Amanoul, S.V., Abdulazeez, A.M., Zeebare, D.Q., Ahmed, F.Y.H.: Intrusion detection systems based on machine learning algorithms. In: 2021 IEEE International Conference on Automatic Control and Intelligent Systems, I2CACIS 2021 – Proceedings, pp. 282–287. Institute of Electrical and Electronics Engineers Inc (2021).
6. Prasad, R., Rohokale, V.: Artificial Intelligence and Machine Learning in Cyber Security. In: Prasad, R. and Rohokale, V. (eds.) Cyber Security: The Lifeline of Information and Communication Technology, pp. 231–247. Springer International Publishing, Cham (2020).
7. Vanin, P., Newe, T., Dhirani, L.L., O'Connell, E., O'Shea, D., Lee, B., Rao, M.: A Study of Network Intrusion Detection Systems Using Artificial Intelligence/Machine Learning (2022).
8. Vasilomanolakis, E., Karuppayah, S., Mühlhäuser, M., Fischer, M.: Taxonomy and Survey of Collaborative Intrusion Detection. ACM Comput. Surv. 47 (2015). https://doi.org/10.1145/2716260
9. Kalimuthan, C., Arokia Renjit, J.: Review on intrusion detection using feature selection with machine learning techniques. In: Materials Today: Proceedings, pp. 3794–3802. Elsevier Ltd (2020).
10. Buczak, A.L., Guven, E.: A Survey of Data Mining and Machine Learning Methods for Cyber Security Intrusion Detection. IEEE Commun. Surv. Tutor. 18(2), 1153–1176 (2016). https://doi.org/10.1109/COMST.2015.2494502.
11. Efe, A., Abaci, İ.N.: Comparison of the Host Based Intrusion Detection Systems and Network Based Intrusion Detection Systems. Celal Bayar Üniversitesi Fen Bilim. Derg. 18, 23–32 (2022). https://doi.org/10.18466/cbayarfbe.832533

12. Mchugh, J.: Testing Intrusion Detection Systems. ACM Trans. Inf. Syst. Secur. 3, 262–294 (2000). https://doi.org/10.1145/382912.382923
13. Anthi, E., Williams, L., Javed, A., Burnap, P.: Hardening Machine Learning Denial of Service (DoS) Defences against Adversarial Attacks in IoT Smart Home Networks. Comput. Secur. 108 (2021). https://doi.org/10.1016/j.cose.2021.102352
14. Jmila, H., Khedher, M.I.: Adversarial Machine Learning for Network Intrusion Detection: A Comparative Study. Comput. Netw. 214 (2022). https://doi.org/10.1016/j.comnet.2022.109073
15. Lippmann, R., Haines, J.W., Fried, D.J., Korba, J., Das, K.: The 1999 DARPA Off-Line Intrusion Detection Evaluation. Comput. Netw. 34(4), 579–595 (2000). https://doi.org/https://doi.org/10.1016/S1389-1286(00)00139-0
16. Lippmann, R., Cunningham, R.K.: "Guide to Creating Stealthy Attacks for the 1999 DARPA Off-Line Intrusion Detection Evaluation", MIT Lincoln Laboratory (2000).
17. Hossain, M.A., Islam, M.S.: Ensuring Network Security with a Robust Intrusion Detection System Using Ensemble-Based Machine Learning. Array. 19 (2023). https://doi.org/10.1016/j.array.2023.100306
18. Cemerlic, A., Yang, L., Kizza, J.M.: "Network Intrusion Detection Based on Bayesian Networks." International Conference on Software Engineering and Knowledge Engineering (2008).
19. Mahdavi, E., Fanian, A., Mirzaei, A., Taghiyarrenani, Z.: ITL-IDS: Incremental Transfer Learning for Intrusion Detection Systems. Knowl-Based Syst. 253 (2022). https://doi.org/10.1016/j.knosys.2022.109542
20. Al-Daweri, M.S., Ariffin, K.A.Z., Abdullah, S., Senan, M.F.E.M.: An Analysis of the KDD99 and UNSW-NB15 Datasets for the Intrusion Detection System. Symmetry (Basel). 12, 1–32 (2020). https://doi.org/10.3390/sym12101666
21. Jayaraj, R., Pushpalatha, A., Sangeetha, K., Kamaleshwar, T., Udhaya Shree, S., Damodaran, D.: Intrusion Detection Based on Phishing Detection with Machine Learning. Meas. Sensors. 31 (2024). https://doi.org/10.1016/j.measen.2023.101003
22. Kilincer, I.F., Ertam, F., Sengur, A.: Machine Learning Methods for Cyber Security Intrusion Detection: Datasets and Comparative Study. Comput. Netw. 188 (2021). https://doi.org/10.1016/j.comnet.2021.107840
23. Rathore, S., Park, J.H.: Semi-Supervised Learning Based Distributed Attack Detection Framework for IoT. Appl. Soft Comput. 72, 79–89 (2018). https://doi.org/10.1016/J.ASOC.2018.05.049
24. Ahmad, M., Riaz, Q., Zeeshan, M., Tahir, H., Haider, S.A., Khan, M.S.: Intrusion Detection in Internet of Things Using Supervised Machine Learning Based on Application and Transport Layer Features Using UNSW-NB15 Data-Set. EURASIP J. Wirel. Commun. Netw. 2021 2021:1. 2021(1), 1–23 (2021). https://doi.org/10.1186/S13638-021-01893-8
25. Saran, N., Kesswani, N.: A Comparative Study of Supervised Machine Learning Classifiers for Intrusion Detection in Internet of Things. Procedia Comput. Sci. 2049–2057. Elsevier B.V. (2022). https://doi.org/10.1016/j.procs.2023.01.181
26. Zhang, H.: The optimality of naive Bayes. Proceedings of the Seventeenth International Florida Artificial Intelligence Research Society Conference, FLAIRS 2004. 2 (2004).
27. Hodo, E., Bellekens, X., Hamilton, A., Tachtatzis, C., Atkinson, R.: Shallow and Deep Networks Intrusion Detection System: A Taxonomy and Survey (2017). https://doi.org/10.48550/arXiv.1701.02145
28. Sahu, S., Mehtre, B.M.: Network intrusion detection system using J48 decision tree. In: 2015 International Conference on Advances in Computing, Communications and Informatics, ICACCI 2015, pp. 2023–2026 (2015). https://doi.org/10.1109/ICACCI.2015.7275914

29. Amin Aburomman, A., Bin Ibne Reaz, M.: A Novel SVM-kNN-PSO Ensemble Method for Intrusion Detection System. Appl. Soft Comput. 38, 360–372 (2016). https://doi.org/10.1016/j.asoc.2015.10.011
30. Santoso, N., Wibowo, W., Himawati, H.: Integration of Synthetic Minority Oversampling Technique for Imbalanced Class – Scientific Figure on ResearchGate. Available from: https://www.researchgate.net/figure/The-maximum-margin-hyperplane-of-SVM_fig1_330168470 [accessed 25 April 2024].
31. Ozkan-Okay, M., Samet, R., Aslan, Ö., Gupta, D.: A Comprehensive Systematic Literature Review on Intrusion Detection Systems. IEEE Access. 9, 157727–157760 (2021). https://doi.org/10.1109/ACCESS.2021.3129336
33. Husák, M., Jirsik, T., Yang, S.: SoK: Contemporary issues and challenges to enable cyber situational awareness for network security. In Proceedings of the 15th International Conference on Availability, Reliability and Security (ARES '20). Association for Computing Machinery, New York, NY, USA, Article 2, 1–10 (2020). https://doi.org/10.1145/3407023.3407062
34. Apruzzese, G., Colajanni, M., Ferretti, L., Guido, A., Marchetti, M.: On the effectiveness of machine and deep learning for cyber security. In: International Conference on Cyber Conflict, CYCON, pp. 371–389. NATO CCD COE Publications (2018).

Chapter 4

Agile security and compliance integration

Enhancing cyber resilience through dynamic, automated processes

Mounia Zaydi, Yassine Maleh, Hayat Zaydi, Youness Khourdifi, Bouchaib Nassereddine, and Zohra Bakouri

4.1 INTRODUCTION

Agile methodologies and innovation in DevOps like implementations offer many advantages to introducing new security controls and automating existing manual security processes and shifts. Primarily, the idea to retrofit security into information technology (IT) in an incremental and self-checking manner makes it an ideal candidate for achieving security objectives. An illustrative comparison of development models, highlighting the shift from traditional to secure agile methodologies, is provided in Figure 4.1. However, agile introduces its own challenges, and security is often seen as a blocker to agility, with secretive risk-averse security cultures being at odds with fast-paced, open, and collaborative agile cultures. There is a risk that agile may forsake security in meeting its quest for speed. There is also a general lack of understanding of how to do security in an agile manner and a lack of tooling to support it. This can result in an increase in technical debt and a negative impact on an organization's security posture. Security innovation in an agile setting is required to ensure that security keeps pace with the rate of change in the business and IT landscape [1–4].

As the number of incidents increases and diversifies, the cost and complexity of mounting a defense against them soar. Approaches to cybersecurity are currently far too slow and reactive to keep up with the pace of change in the cyber threat landscape; they are also unmanageable. Some of the attacks that have really been widely mentioned recently, in fact, have also translated to new regulations and legislation, particularly the new requirement of automating security controls to ensure the organization's security posture remains in a known good state for the minimization of cyber risk exposure. This kind of fundamental shift represents a tipping point for security practice—this will fundamentally change security models and processes. Nonetheless, many new security controls and processes cannot be retrofitted onto the existing IT systems, never designed with security in mind. High manual intervention security practices that have relied upon are no longer affordable or scalable,

DOI: 10.1201/9781003478676-4

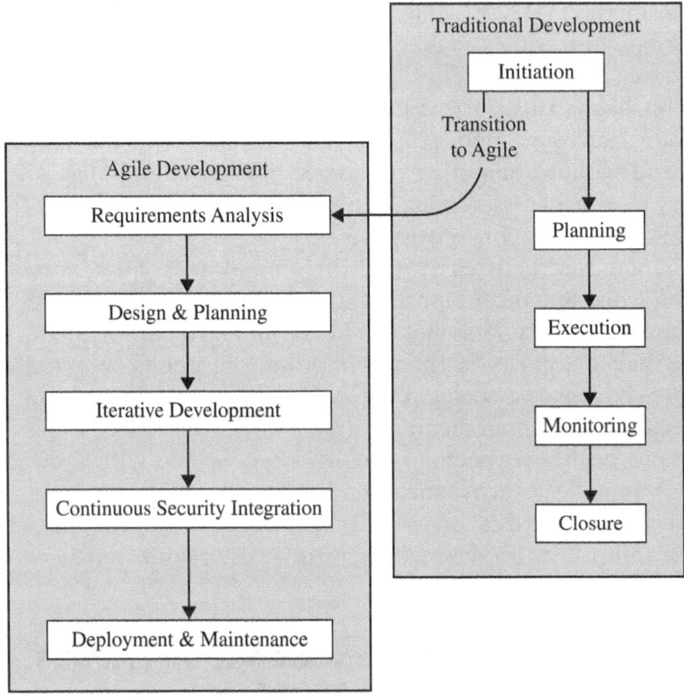

Figure 4.1 Comparison of development models: Traditional vs. secure agile.

and a complete overhaul of security processes and controls is required [5–8]. Meanwhile, rapid cyber threats continue to escalate without reprieve. In fact, as late as 2017, the Department for Digital, Culture, Media and Sport reported that 46% of all UK businesses had identified at least one breach or attack. Security has often been an afterthought or has been something that just didn't move the needle enough to integrate well with the pace of agile development and agility. The fast-changing nature of technologies and platforms makes sure traditional security practices are now largely incompatible and unable to deliver suitable protection to the current IT systems and practices. With organizations rapidly moving to DevSecOps methodologies or similar and a move to cloud platforms with high utilization of microservices architecture, security will have to evolve continually to cope with the industry and still be an enabler for change [8–10]. Traditional security measures built security toward the end of the Software Development Life Cycle (SDLC). This, though, goes against the very nature of agile practices wherein, due to development and operations looking to match the speed and development curve of the business, security is often seen as deprioritized. Traditional security measures, such as weeks of waiting time and repetitive security processes for literally every change, can inhibit the capability for rapid change that sets agile apart from traditional methods [11, 12]. The last ten years or so have seen a sea change in the way organizations worked when their IT infrastructure moved from the traditional methodologies of waterfall development to practices driven by agility. This has in turn further complicated efforts to develop products and services at a faster and more efficient rate because of the increased adoption of cloud and mobile computing platforms. Meanwhile, despite all the value this agility can bring, its acceleration and iterative nature have often seen a significant compromise in security [13–15].

This chapter is structured around three main axes. First, we explore the integration of compliance requirements into the development process, highlighting how agile methodologies can be harmonized with regulatory imperatives to enhance security without compromising agility. Next, we address the specific challenges associated with compliance in agile environments and the strategies to overcome them, illustrated with concrete examples from the financial and healthcare sectors. Finally, we conclude with a discussion on automation in agile security, shedding light on essential tools and their benefits, as well as successful case studies that demonstrate the impact of these integrated approaches on the security posture of organizations.

4.2 INCORPORATING COMPLIANCE REQUIREMENTS INTO THE DEVELOPMENT PROCESS

It becomes feasible to assign features or tales to controls according to their risk value by utilizing predictive analysis in the risk management matrix. This enables the dynamic prioritization of control implementation across several projects, ensuring that business-critical controls are consistently integrated. Specific control types can be applied to features, and the system can be configured to filter and show those features that are connected to calling up a group of control statements. This will create a precedent for the development of functional system requirements and aid in bridging the knowledge gap over what is needed to meet controls. Modifications to control will result in modifications to access to features, which will then cause another modification to access to coded logic. It is possible to declare that a certain control has been fully implemented at this point. Finally, a meta-model that can be validated and questioned utilizing logic written within an expert system can be created by describing control or control type specific metadata. As a result, complicated compliance assessment and assurance tasks will be able to be automated [16, 17]. Conventionally speaking, compliance assessment happens at project completion, when the system is approved for operation and the corresponding controls are confirmed to be in good working order. As a result, it is neglected during development and frequently necessitates extensive revision because rules cannot be satisfied without careful design attention. Penalties for nonconforming systems and the cost of delays incurred due to unmet controls provide a comparable picture of the cost of this compliance gap. Compliance needs to be considered during the development process to guarantee no loss of development or additional expense. It is possible to employ automation to make sure that control needs are met utilizing agile approaches by integrating compliance into the development process. Since automation usually refers to the use of code or tools to replace human activity, it is frequently perceived as being in opposition to agile. Though this idea can be best applied to lower the cost of compliance, it is more appropriate to think of automation as an acceleration of work completion. It is feasible to make sure that systems created in a short amount of time do not get around control needs by automating the process of checking a control or putting in place a specific access rule. Given that access to control change is far more expensive, it is possible to guarantee that there is no delay encountered at the end of the project if agile techniques can be stopped with each specific control being implemented. Overall, the outcome is a system that satisfies security and compliance objectives and further confidence that no control needs are disregarded [18–20].

4.2.1 Challenges in agile compliance

Capability Maturity Model (CMM) has long been favored by managers and engineers to evaluate software processes within their organizations and implement improvements. These improvements, in turn, lead to time and cost savings for projects. Recently, there has been a development in CMM called flexible CMM, which is essentially a theoretical concept proposing a systematic and well-planned approach to change. While there may be various interpretations of this flexible change, we will specifically explore the changes suggested by CMM for software process improvement. When considering the integration of CMM with an agile method, a practitioner has recommended, "map CMM2 to eXtreme Programming methodology (XP). Cette méthode est l'une des plus connues dans le cadre des approches agiles pour le développement de logiciels. Voici quelques principes et pratiques clés de XP qui illustrent son app XP, and CMM3 to XP, but not to XP before we get to that." This statement clearly indicates that rigidly applying CMM is not suitable for agile software processes. Instead, the focus is on understanding the relevant statutory and regulatory requirements. The practitioner's remark about not moving to XP before reaching a certain point demonstrates the importance of a gradual progression through the levels of both XP and CMM during software development. At this stage, an agile team is on the verge of transitioning into a simplified version of CMM, which differs from its current state, to meet the target level for the CMM process. This example highlights the potential benefits of making change decisions that lead to a more advantageous approach. When striving to improve a software process, it is common to compare the new process with the old. However, it is crucial for the team to concentrate on remapping a CMM level before implementation begins. Otherwise, the considerations and requirements associated with the CMM level may gradually fade as the team becomes engrossed in the activities of the new process. This can make it challenging to adhere to specific requirements and effectively implement change when the team is already engaged in an existing activity. This method often contrasts with a bottom-up approach, where the focus is on the nature of activities and the objectives necessary to achieve a new process [21].

In agile development, change is professionally accepted. Teams at numerous organizations have successfully implemented agile development methodology for their software development projects. Agile development requires a higher degree of flexibility to execute changes many times without affecting the cost and schedule of the project. Agility in agile development is achieved through improvement in task-worker relation by implementing professional best practices [22–25].

4.2.2 Strategic integration

One of the main objectives of research in this area is to make a scientific contribution to improving software security by means of a new advance in the same field. The research aims to demonstrate the effectiveness of an empirical method, through a case study, in achieving self-executing compliance [26, 27].

In general, the applied method of research is the improvement of agile development and compliance with case study research. The expected output is an implemented software with an embedded compliance. Hence, the research focuses on seeing the difference in activity between the previous agile development and compliance with this improved method through a comparison of both activities, and what the positive impact is from the improvement [28, 29] (Figure 4.2).

Regarding the compliance process activity within an agile development environment, it is like software production, which is based on the plan established and the requirements gathered at the beginning of the project. Software production is divided into several development phases, and the software is tested to ensure it functions and meets expectations. The goal of this research is to enhance compliance in secure and agile development. The outcome of compliance integrated into software should be more effective than manual compliance. The steps of integrated compliance should improve software security and minimize the risk of security attacks. The positive outcome is a contribution to enhancing software security and minimizing security attacks [30–32].

4.2.3 Practical examples

The finance sector company selected for the case study is a large UK-based financial organization with operations in investment banking, retail banking, and insurance. Due to the nature of their industry and the significance of the data they handle, they have many compliance regulations to adhere to. In particular, the Sarbanes-Oxley Act (SOX) has significant sections about the need for internal controls and reporting on the effectiveness of those controls

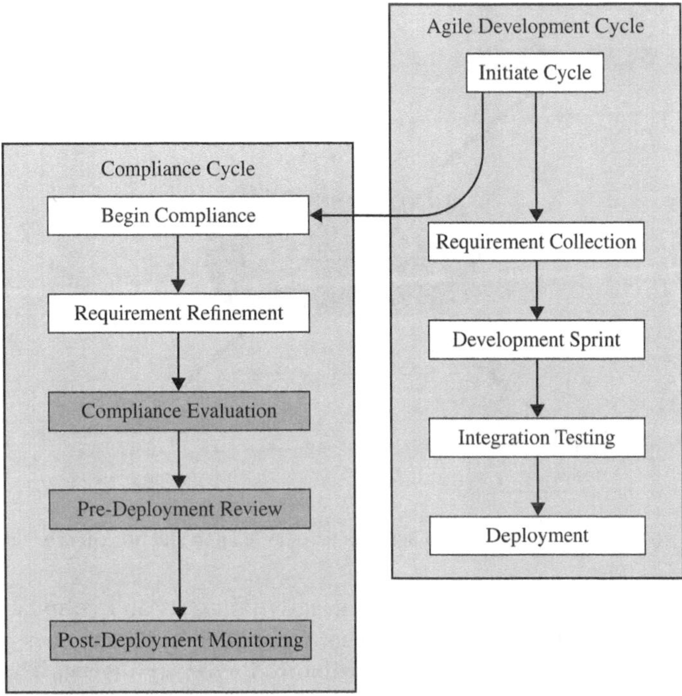

Figure 4.2 Integrating compliance requirements into agile development.

for public companies. Many finance companies like the one in this case study have looked at this sort of regulation and been put off attempting to use agile methods because they feel they are incompatible with the imposition of heavy control. This company engaged the author to provide training and consulting around agile methods with the long-term aim of enabling their development teams to be more effective and deliver software of higher quality. During this engagement, as an experiment, one team decided to take what was then a fledgling agile framework and attempt to implement this method of software development to create an internal reporting system that would pass an SOX audit. This case traces the methods employed by that team and the agile and compliant hybrid framework that emerged as a result [33, 34]. Real-world applications of this integration are demonstrated in Figure 4.3.

4.2.3.1 Case study 1

In the finance sector, a leading global investment bank successfully integrated compliance measures within their agile framework, resulting in enhanced cyber resilience. Similarly, in the healthcare sector, a prominent hospital implemented agile practices and seamlessly integrated compliance protocols, leading to significant improvements in cyber resilience and data security. The

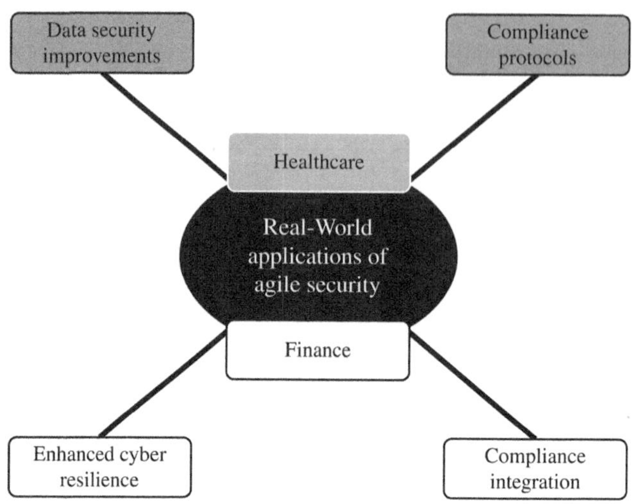

Figure 4.3 Real-world applications of agile security in finance and healthcare.

hospital's successful implementation of agile practices and compliance integration serves as a practical example that highlights the effectiveness of bridging compliance and automation for enhanced cyber resilience. The finance industry case study demonstrates the positive outcomes of incorporating compliance measures within agile frameworks, while the healthcare sector example further reinforces the importance of this integration in bolstering cyber resilience and safeguarding sensitive data. These case studies from the finance and healthcare sectors provide compelling evidence of the benefits of bridging compliance and automation for enhanced cyber resilience, emphasizing the need for other industries to adopt similar strategies. By showcasing successful examples in both sectors, this section aims to inspire organizations in different industries to explore the possibilities of integrating compliance measures within their agile frameworks for improved cyber resilience. These case studies from the finance and healthcare sectors provide a roadmap for organizations in various industries to navigate the challenges of integrating compliance measures within their agile frameworks, ultimately enhancing their cyber resilience capabilities. By examining these real-world examples, organizations can gain valuable insights and practical strategies for effectively bridging compliance and automation, ultimately improving their cyber resilience, and minimizing potential vulnerabilities. These case studies from the finance and healthcare sectors provide tangible evidence of the tangible benefits of integrating compliance measures within agile frameworks, showcasing how organizations can enhance their cyber resilience and mitigate potential vulnerabilities. By analyzing these case studies, organizations can gain valuable insights on how to successfully integrate compliance measures within their agile frameworks, ultimately improving their cyber resilience and fortifying their defenses against potential cyber threats. These case studies from the finance and healthcare sectors offer concrete evidence of the effectiveness of bridging compliance and automation in enhancing cyber resilience, providing organizations with valuable insights on successfully integrating compliance measures within their agile frameworks and strengthening their defenses against potential cyber threats.

4.2.3.2 Case study 2

In the healthcare sector, a prominent hospital system demonstrated the effective integration of agile methodologies to enhance compliance and security within its patient data management systems, focusing on Health Insurance Portability and Accountability Act (HIPAA) regulations. Adopting a scrum framework, the project involved cross-functional teams of software developers, compliance officers, and IT security experts who employed iterative development cycles for rapid updates and frequent compliance reassessments. The integration of automated compliance checks within the continuous integration/continuous delivery (CI/CD) pipeline streamlined validation processes, reducing manual efforts, and accelerating deployment. Despite initial resistance due to the perceived incompatibility between agile speed and rigorous compliance reviews, targeted training and automated tools facilitated a better understanding and implementation of agile methodologies in this regulated environment. This approach not only improved the hospital's IT system security and compliance but also set a benchmark for deploying agile practices effectively in environments with stringent regulatory requirements, thus enhancing overall cyber resilience.

4.3 AUTOMATION IN AGILE SECURITY

Information system security automation and agile development is a relatively unexplored research area. The goal of this topic is to provide an understanding of agile development and how it can be related to security automation. Agile development methods have been created to increase code quality, decrease development time, and increase adaptability to changing information system requirements. These are not traditional waterfall methodologies in which each phase of development is rigid and defined. Agile development increases flexibility into the software development lifecycle. Security has been a latecomer to the software development lifecycle. Information systems are often vulnerable in their growing stage, and security has commonly been seen as a detractor to speed of development. This has resulted in huge technical debts in security, in which patching and plugging security holes is a far more difficult and costly task compared to secure development from the beginning. This is an area of interest to which findings in agile development and its relation to security automation can be a great benefit to today's security professionals [25, 35, 36]. An illustrative example of how compliance automation

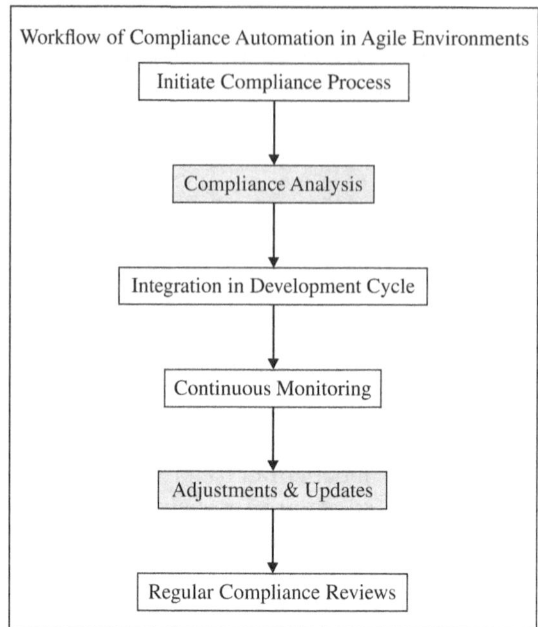

Figure 4.4 Schematic representation of compliance automation processes within agile frameworks.

can be integrated within agile environments is shown in Figure 4.4. Providing a knowledge of agile development and its potential relationship to security automation is the aim of this topic. Agile development methodologies aim to improve code quality, reduce development time, and boost flexibility in response to evolving information system needs.

4.3.1 Benefits of automation

Comparison from a Computer World article included a real account of how automation saved money and time in addition to statistical averages [39]. The author initially discussed how much time and money departments spend administering and repairing manually hacked systems using two distinct IT organizations. The author described a scenario in which two businesses with Windows systems susceptible to Structured Query Language (SQL) Slammer were chosen at random to receive an attack a few weeks later. It only took the IT staff of the first organization a few days after the patch was issued to take the proactive approach and use automation to correct the vulnerability. It took them only 15 minutes to make sure all their systems were fixed. The IT department of the second firm, which did not use automation, spent 100 hours applying the patch to 350 different systems since they were unsure if their systems were patched. After a span of two weeks, the system administrators of the second organization were compelled to interrupt their daily schedules to rectify the compromised systems through the reinstallation of Windows and different software. A total of 69 hours of lost productivity and $17,250 were incurred by the administrators due to the 29 hours spent making sure the systems were patched correctly and the 40 hours spent patching the systems. This $17,250 was but a tiny portion of the overall harm caused by the Slammer virus. The Damage Assessment Teams of the two organizations conducted research that indicated the infection cost the entire organization more than $5 million in IT work. The author then compared the time and resources used by the first company which had automation to the second which did not step-by-step. The author ended with a statistical overview of a study conducted by IT administrators, indicating that fixing an issue after it has already happened is 20%–40% more expensive than preventing a problem in the first place. Returning to the scenario, the author spent a few pages discussing how modern IT professionals' high-priority lists are growing longer and longer due to the increasing number of IT Security Threats or Problems. This job is typically scheduled later, and in the absence of automation, it is frequently stopped or causes more serious, time-consuming issues. The author concluded by stating that it is certain to improve the quality and efficacy of adopting a security solution if resources are freed up and the time it takes to fix or install one is further shortened [37–39].

4.3.2 Enhancing agile development security through automated analysis techniques

An essential requirement for a modern CI/CD pipeline is integration support. With strong integration, the tools can automate a large section of the development cycle. Static analysis tools usually have a very steep initial learning curve and require extensive configuration. Therefore, it's important that integration doesn't necessitate a lot of script writing and custom configurations. Dynamic analysis testing can be performed by creating a build in the CI cycle and applying testing to the created binary. The full analysis will provide a report on the entire build and can be achieved with a minimal amount of configuration. In a CI cycle, it's possible to automate dynamic analysis such that the report is readily available for analysis by developers in the morning. This is vastly different from traditional security testing that is normally performed toward the end of the development cycle. Static analysis and dynamic analysis are best performed after each setting is checked. This prevents an accumulation of errors and latency in fixing issues [40]. Static analysis is a very powerful tool in identifying software bugs and security vulnerabilities that are hard to find using functional testing. Static analysis tools can be very customizable and allow an organization to set a specific standard of security quality for source code. These tools will display a list of violations, with the importance and proper location of each violation found. The developers can then go through and fix these violations, which in turn improves code security and quality.

Static analysis can be performed as a git hook before the code is committed, before a merge, or after a build. This is advantageous because it allows the user to fail fast on a build and avoid preferring off fixing an accumulation of violations later [41].

Many organizations have had to find a trade-off between security and rapid deployment. There are specially designed security-focused CI/CD pipelines called DevSecOps and Runnable that have been designed to automate security testing for code. These tools differ from traditional CI/CD pipelines because they are designed to enforce security policies that terminate the launch of builds known to contain security vulnerabilities. They do this by scanning and testing the source code using dynamic and static analysis tools. These tools are designed to make life harder for developers to commit code if it is not of good security quality—MISRA C and ISO/IEC 17952 compliance. This may seem counterproductive to the initial goal of a CI/CD pipeline, but it is the correct and desired behavior because it's preventing the deployment of unsafe software [42]. The CI/CD pipelines have now become an integral part of the software development process. They are designed to save time and effort in software development by taking coding from a developer and then building the software, testing it, and deploying it by the click of a button. This may seem too good to be true, and it is. The problem with traditional CI/CD pipelines is that they focus too much on rapid deployment and lack the built-in security checks to test and verify if the code that is being deployed is safe in the first place. The source code that is being deployed may contain several exploits or vulnerabilities and the rapid deployment systems don't do any kind of checking to ensure that the software that is going to be deployed from the CI/CD pipeline is safe. This just makes life that much easier for an attacker who can easily find and exploit vulnerabilities within your source code [43] (Figure 4.5).

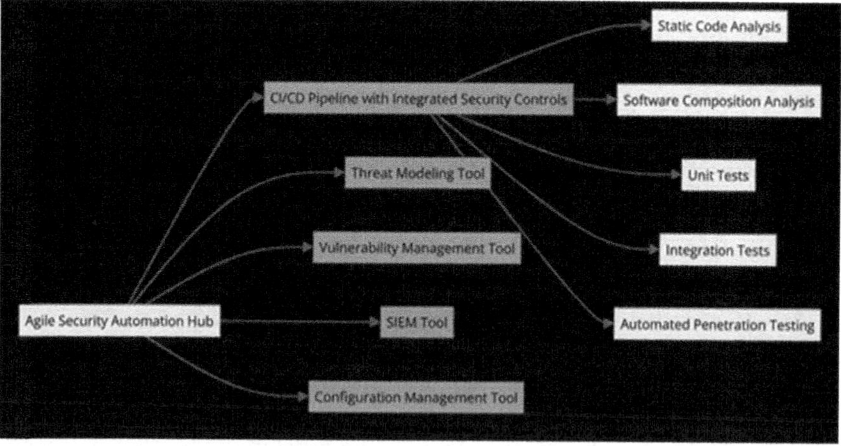

Figure 4.5 Automation tools for agile security.

4.4 THE SYNERGY OF DevOps, SecOps, AND DevSecOps IN AGILE SECURITY MANAGEMENT

The challenge is to consider security as a reflection of the aggregated requirements of business and operational IT systems, with the ability to apply and modify security policies in near real time [44–46]. For this to work, a holistic understanding of the entire development lifecycle is required, and therein lies another problem. Security teams often don't have the resources to keep up with the pace of today's agile development environment. We need security technologies and a community of people and processes to be more agile. This means minimizing on-premises infrastructure security and leveraging agile policy-based security automation, including security functionality that can be used directly by development and operations IT teams [28, 47]. However, industry leaders have been energized by the success of their agile development programs and want security to engage with and follow, if not surpass, agility. Accepting that it means different things to different people, the heart of agility emanates from the ability to initiate happy change on pervasive issues. Thus, agile security focuses not only on implementing security measures as part of agile development but also on transforming security into an agile process, using the overall risk management approach of ISO 27001 as a model. This involves synthesizing threat information and risk analysis into security requirements, then translating these into risk-aware security policies to better protect the organization [48]. Initially, the terms "agile development" and "security management" were not seen as things that could work in unison. Marketing, operations, and development embodied agility, while security seemed to be a delaying factor: cumbersome policies, strict deadlines, and an inflexible attitude [11, 28].

4.4.1 Conceptual overview

There is no conflict between DevOps and SecOps. Given that the terms "DevOps" and "SecOps" are in the same title, you might think that they are synonymous. Development and operations are combined to form DevOps, after all. There is nothing that is falser than this. Though they come from the same root, DevOps is a "software engineering culture and practice," whereas Ops is primarily concerned with "keeping the lights on." Stated differently, SecOps is about guaranteeing the security of what was developed, whereas DevOps is about producing things, including on the operations side like deployment frameworks and automation scripts. This is not to suggest that the two cannot coexist. The goal of DevOps' inception was to close the knowledge gap that existed between security and other IT disciplines, such as development. Building on the agile and lean approaches, one of the main objectives was to promote a cultural shift toward a high-trust, cross-functional team that iterates quickly to produce high-quality software. Though it does not specify security's specific role, it is a positive step. After taking a few more positive moves, this developed into what I like to refer to as "Security Theatre." This happens when security protocols within an organization are not genuinely intended to be secure but are instead unduly focused on conforming to security standards or appearing that way. This involves treating security as a separate post-development phase, or worse, conducting security infrequently at the conclusion of development cycles. Regretfully, a lot of organizations still fit this description today.

4.4.2 Integration strategies

The key safety issue in agile at an organizational level is the ability to effectively integrate safety practices. This allows the incremental nature of agile development to continue unhindered, with all the benefits of agile intact, while maintaining safety. Integrating security practices within a budget or schedule is already a major problem in software development. Considering security as it is incrementally implemented is an additional step. It involves understanding where security threats arise in the development process and ensuring that security practices are in place to counter these specific threats. This means reflecting the incremental development process, with security practices adapted to each incremental stage. For example, unit testing is a deeply ingrained practice among software developers; for security to be tested incrementally, methods must be used that are directly adapted to each development activity. To achieve this, security specialists will need to have an in-depth knowledge of development processes, while developers will need to have greater expertise in security. This is a difficult requirement and automating security checks and tests wherever possible is essential to make this a cost-effective possibility [49–51].

4.4.3 Case studies of effective DevSecOps deployments: Enhancing organizational security postures through agile integration

The DevSecOps integration framework, which is key to understanding how security practices are integrated into agile processes, is illustrated in Figure 4.6. We distinguish many success stories that demonstrate the effectiveness of integrating security into agile development frameworks. Unlike stories that focus solely on failures and disasters, these stories show how proactive safety integration can lead to positive change. A particularly noteworthy case is that of the financial sector. High-Tech Bridge, a global provider of web security testing, helped a major financial organization dramatically improve its ability to identify vulnerabilities in web applications and software. By conducting more than 50 security tests and simulated attacks in different countries, the organization was able to significantly reduce the likelihood of data loss due to web-related software vulnerabilities, while achieving substantial cost savings. During these tests, an interesting correlation was established between the probability of a successful exploit by a hacker and the programming language used to develop the software. This finding led to the development of a cost-effective training program for developers, aimed at reducing the risk

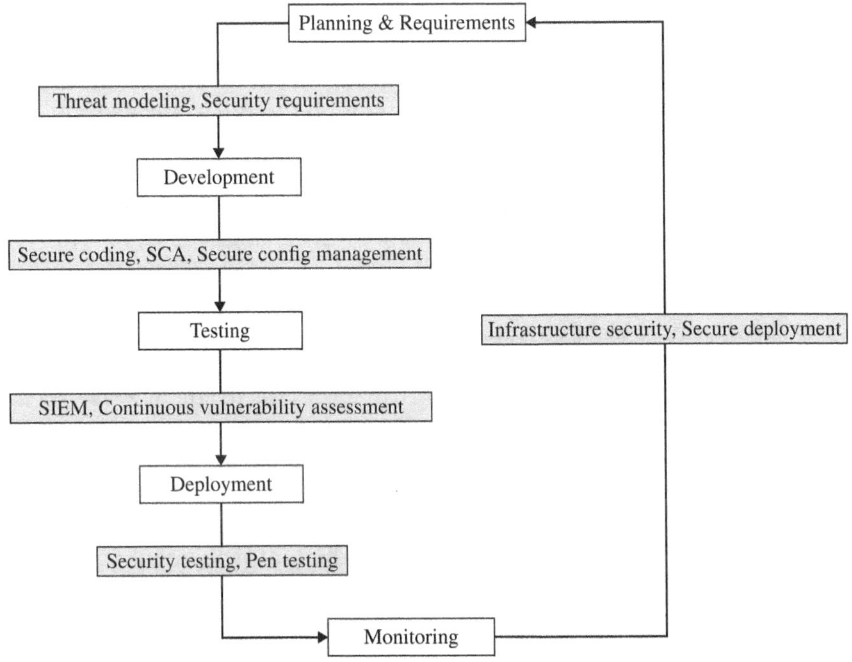

Figure 4.6 DevSecOps integration framework.

of vulnerabilities. This example highlights the impact of the right strategic approach and targeted training to enhance security at low cost [52, 53].

4.4.4 From reactive to proactive: How industries are fortifying cybersecurity

The primary objective of any industry is to better secure and protect information assets and data from potential loss or exposure. The focus on securing software systems through secure software development cycles and increased awareness of information sharing has seen secure coding become the most under-utilized high-impact security initiative. Encryption has gained momentum and data loss prevention (DLP) has moved into the top five. The finance and insurance sector has always been a hotspot for malicious attacks or data breaches motivated by greed, but it appears to have improved this year with a significant reduction in incidents, with the short-term aim being to support this long-term trend. The public sector still faces challenges, but it is encouraging to see the focus on incident preparedness and response [54, 55].

4.4.4.1 Industry segmentation

Data from the most recent Verizon Data Breach Investigations Report (DBIR) shows a shift in security philosophy and operations within organizations in response to the ever-changing security landscape. Faced with the risks associated with rapid exposure of their data, it is no wonder that businesses invest time in tracking and preventing incidents. This year's report sees an increased focus on making data more secure, by either enacting policies or implementing security solutions, with a notably marginal decrease in incidents. The shift in focus has seen an increase in data protection activities such as data masking, encryption, and redaction, while antivirus remains the most widely deployed solution, especially in organizations with less than 1000 employees. Let's take a detailed look at the changes observed this year across various business sectors [56–58].

The data indicates a marked shift toward more robust security protocols across all sectors, with a particularly noticeable emphasis in the healthcare and financial industries. In the healthcare sector, there has been a concerted effort to bolster data privacy measures in response to an increase in targeted cyberattacks, which often aim to exploit sensitive patient information. Organizations within this sector have accelerated the deployment of advanced encryption techniques and have significantly expanded their use of data access controls to limit exposure to unauthorized entities.

In the financial sector, firms have increased their investment in threat detection and response capabilities. This includes the adoption of more sophisticated monitoring tools that can detect anomalies in transaction data, which is often a precursor to fraudulent activity. Moreover, there has been a significant rise in the use of artificial intelligence (AI) to automate the detection of patterns indicative of cyber threats, thereby enabling quicker responses and reducing the window of opportunity for attackers. The retail and e-commerce sectors have also seen considerable changes, primarily driven by the need to secure online transactions and protect consumer data. Enhanced security measures such as multi-factor authentication (MFA) and secure payment gateways have become more prevalent, aimed at thwarting phishing attacks and securing online payment processes. Overall, these changes reflect a growing recognition of the critical importance of cybersecurity within the business environment. As cyber threats continue to evolve in complexity and scale, the proactive measures taken by these sectors aim not only to protect their own assets but also to safeguard the interests of their customers and stakeholders. The ongoing adaptations and enhancements in security practices are crucial to maintaining trust and ensuring the resilience of business operations against the backdrop of an increasingly hostile cyber landscape.

4.5 LITERATURE REVIEW: CHALLENGES AND TRENDS IN HEALTHCARE SECURITY, PRIVACY, AND REGULATORY COMPLIANCE

Literature on healthcare illustrates the importance of meeting security and privacy requirements in compliance statutes, and the difficulty in doing so. In a study of Food and Drug Administration (FDA) regulation of medical devices, Harrison et al. identify a trend toward more specific regulation of software in medical devices and state the software industry is unprepared to deal with it. They note a tension between the FDA's intent to ensure safety without stifling innovation and a lack of clear methods for ascertaining safety of software relative to risk. This is very similar to the trade-off of assuring security without sacrificing functionality and is complicated by the fact that software risk can be highly domain specific. FDA regulatory requirements are detailed and have the force of law and therefore are essentially compliance requirements. In a separate study of privacy and security regulation in medical informatics, Goodman et al. identify a complex web of laws and requirements which create a high liability environment. They describe the HIPAA Privacy Rule as more stringent than HIPAA Security Rule and as imposing an unrealistic set of privacy protection regulations. While this may be more an issue of HIPAA rule implementation, it does suggest a gap between requirements and feasible activity for the laws. In a recent paper, Malin et al. provide empirical evidence that confusion about privacy regulations can prevent appropriate data access and uses a case example of requirements conflicts. An interesting contrast to all this comes from a recent study of informal work practices in the healthcare sector, where there has been little consideration of compliance and security requirements. This type of domain and requirement analysis work would be highly beneficial to target sector [59].

There are a broad range of academic subjects which underpin computer security and privacy, including protocol design, databases, distributed systems, cryptography, AI, and many others. A key impetus for this work is to better understand the complex and sometimes conflicting interplay of security and privacy requirements with other functional requirements. This means understanding the requirements themselves, and how they are represented in domains and with different stakeholders. Our methodology work is directly aimed at bridging the compliance and security spaces, which are interwoven with privacy concerns. The work reviewed includes a broad range of empirical studies, methods, and system developments. While a full survey of all relevant work in security, privacy, and compliance is beyond the scope of this chapter, we review enough to identify the key common themes and challenges, and to evaluate our methodologies and systems in that context [60–62].

4.6 THE PROPOSED AGILE ADAPTIVE SECURITY MODEL (2ASM)

It is now time to consider the design of security processes by seeking the best possible fit with agile practices while maintaining the necessary control environment. An agile security process model should feature attributes that reconcile the requirements for agility and control. An approach is needed that allows for rapid changes and responses in the security process, where changes are inherently safe and provide necessary information for audit and compliance purposes. Actions and reactions in security changes must enable swift transformations within the business. Often, changes in security requirements have been perceived as a barrier to business progression. Security is seen as blocking business activity by enforcing mandatory requirements or implementing controls that halt certain operations. In today's business world, it is unacceptable to block activity, and changes will often prompt modifications in security requirements. Issues arise when security changes lag business changes, leading to cases of non-compliance and system exposure because security controls were not applied in a timely manner. Changes are often based on trial and error, which is the very essence of agility in security. A modern idea or principle that could specifically be employed to approve the implementation of security requirements relative to activities and changes would involve redefining the activity based on certain risk criteria and establishing a cost-benefit analysis of applying the security control or modifying the security requirement. Activity changes will then be reassessed, and requirement changes can be reflected at the appropriate time. This proposal is still an abstraction, and a more detailed mapping with specific agile methodologies will provide a practical means of implementing and adding a level of agility to the security process. Considering agility is still necessary to assert that changes in security requirements are correct and that iterations do not introduce a higher level of risk to the systems. This is merely an implication. It is obviously desirable that any part of the security process can be revisited if the security control does not match the requirements.

4.6.1 Description of the proposed approach: Integrating security from the earliest stages of development

The integration of security must be based on these security requirements within an iterative process throughout the development stages and should involve a mix of specialized security experts working with the development team. An example could be an application that needs to restrict access to a specific URL based on user roles. The functional requirement would be that only users with specified roles can access the URL. The security requirement would be to ensure that access to this URL is never possible for someone without these roles, even if they have a direct link. A possible design for this would be to use method-level security and intercept all calls to the method defined for this URL with a security check. The conventional approach of adding security in the final stages of development through penetration testing and audits is flawed. In an era of rapid agile development, we need to integrate security from the beginning. Basic defense practices such as using a firewall, antivirus, input validation, and proper session management should still be used when appropriate. However, the focus of the security integration approach is to own the security requirements and implement the necessary practices as both functional and non-functional requirements within the infrastructure and system design.

4.6.2 Model illustration

In the approach we propose, the adaptation and integration of security are envisioned as an evolutionary and iterative process. This process is based on a series of incremental steps aimed at achieving specific intermediate states that add value to the management of security and compliance, regardless of the realization of the final state. This strategy is described as "guided evolution," which avoids radical efforts of top-down reengineering and promotes sustainable modifications to the infrastructure and work practices. The model we define, schematically represented in Figure 4.7, relies on a thorough understanding of the current state of the organization's security and compliance requirements, what is already in place to meet these requirements, and what is necessary to mitigate additional risks presented by changes. Our goal is to define a long-term strategy for automating security and assurance, while emphasizing the need to consider compliance within a dynamic business and IT environment. This method aligns with the philosophy of agile development methodologies, such as DevOps, making it particularly relevant for companies seeking to harmonize their IT security management with agile IT operations. The approach offers a way to progress through successive improvements, not seeking immediate perfection but continuously adapting to the changing needs of the business and emerging external threats.

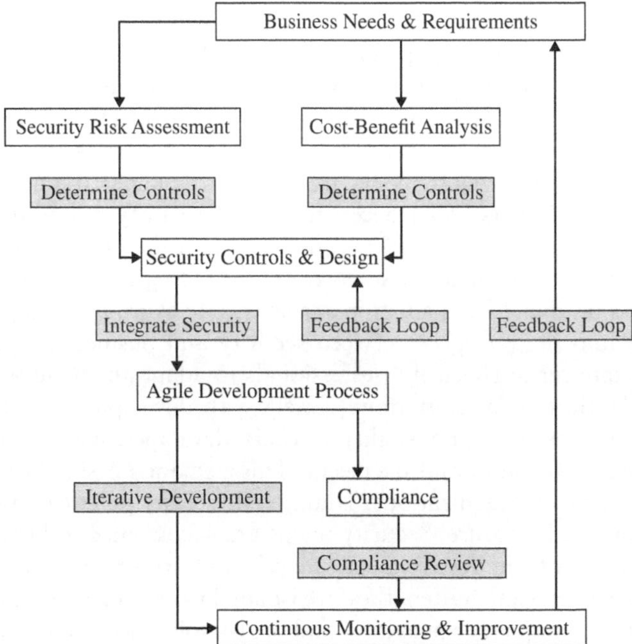

Figure 4.7 Proposed secure agile methodology: Agile adaptive security model (2ASM).

This tight feedback loop is designed to assess the impact of security activities to adjust them and make them more effective. Security activities can identify a need for change in security controls. An example might be the requirements for applying security controls for the confidentiality, integrity, or availability of the data that the application processes. A security control may be new or may require a change to reflect the way data is processed within the application. In some cases, security requirements may not change, but security controls may need to be altered to optimize costs or resource requirements. An example of this could be an impact on the performance of an application resulting from a resource-intensive security control. The close integration of security and development activities allows for rapid and effective changes in security controls. A repeatable effort is continuously integrated into the lifecycle with the grouping of security and risk management practices, based on compliance and policy controls. By adopting this proactive approach, organizations can improve their security posture and ensure that appropriate security controls are effectively implemented. As agile development processes are iterative, security activities must also be iterative. It is within these short iterative cycles that organizations find the opportunity to integrate security activities into their development activities. These shorter cycles allow for a tight feedback loop between security and development activities.

4.6.3 Discussion and challenges

Agility aims to address this through proactive identification and mitigation of security risks within a development iteration. However, security teams may struggle to adapt to an upstream approach. They are required to assume a leadership role by collaborating with the business to identify and understand compliance and risk management objectives and work with development teams to ensure that security requirements are clearly articulated and integrated into functional requirements. This necessitates that security teams deepen their understanding of business operations and grasp how the integration of security and development can impact a software project. Regrettably, many security professionals originate from a Computer Science background and possess limited insight into the interplay between security and business processes. This may necessitate certain security professionals to adapt and acquire new competencies. As they collaborate more closely with development, security must eventually become the responsibility of every developer, with security teams serving more as mentors and overseers. This demands a significant cultural shift within security organizations. Traditional security practices are often tactical and reactive by nature. Security teams are accustomed to being incorporated at the end of the development lifecycle and are provided with a list of security tasks to execute. Often, they are isolated from understanding the business objectives, compliance penalties, and/or risk tolerance and are left to independently determine how these business factors should translate into security policies. This results in security requirements that are ambiguous to developers and are not verified until a security assessment is conducted post-development. When security flaws are identified at this stage, developers are often reluctant to address them as it may cause delays in release. Compared to traditional security implementations, agile security practices represent a significant departure from past approaches. They require a high degree of automation, integration into the development lifecycle, and management of risk and compliance processes. They emphasize the importance of effectiveness and documentation of security activities over adherence to specific procedures. The change imposed by agile methodologies on security teams and the organization presents a new set of challenges that need to be understood and addressed.

To make this vision of the future, several technology and culture shifts must take place. The security industry, to some extent, must accept the probabilistic nature of future digital systems and do away with the ideal that secure systems are those for which there is no possibility of failure. This is not to say that the idea of security as an optimization problem with respect to a well-defined objective function is without merit. However, a probability distribution over possible states for a dynamic system does define a clear set of possible "good states." Adapting security policy and system state to maximize the probability of a state being in that set is a concept which maps cleanly to current security theory, even if the methods by which one might prove that a system satisfies a given property are not yet established.

4.7 CONCLUSION AND FUTURE DIRECTIONS

The concept of a federation of best-of-breed security and compliance management solutions can make integration much more digestible. Each solution is responsible for its own integration using common technologies and design patterns, to produce an integrated view of that solution's domain, and address the connections between that domain and other security and compliance-related business processes and data. Organizations can start small, addressing the most acute security and compliance requirements, and expand integration efforts as they prove value. Value is realized at each step, helping to justify further investment. With integration happening between solutions rather than within a grand monolithic architecture, there is less risk of one integration project impacting others. And if a particular best-of-breed solution does not deliver value, it is easier to replace it with an alternative.

The benefits of implementing an integrated system on security and compliance management are clear: a single, unified view across various IT and compliance silos enables an organization to comprehend their compliance and security postures in context. This creates a platform for rationalizing existing resources, making more effective risk-based decisions and investments, and sharing data across the organization to support more effective security and compliance processes. The challenge is to form an integration strategy that the organization can afford to build and maintain, and that delivers the expected value. In years past, the high cost of integration has led some organizations to adopt an all-or-nothing approach. This requires a massive upfront investment to integrate everything without a clear understanding of the value delivered. Some have been put off by the perception that integration is all pain and no gain.

This chapter has presented the case for embracing automated compliance for cybersecurity. Ultimately, the security industry must keep up with the digital industries it intends to protect. The use of unaltering standards for the implementation of dynamic systems is not sustainable. However, the development of security policy and standards which are capable of dynamic application is a tall order. To impose a static set of constraints on dynamic systems is to invite either circumvention of those constraints or system failure. Dynamically and automatically adapting security policy is a step forward in solving this problem. By making security policy itself a dynamic construct, which can adapt to changes in the system it protects, we remove the requirement for human intervention to maintain a secure state and minimize the probability of unintended exposure due to unenforceable constraints.

REFERENCES

[1] D. Nyale, and S. M. Angolo, "Unravelling DevOps agile methodologies: A comprehensive review of recent research," researchgate.net

[2] M. Poláková-Kersten, and S. Khanagha, "Digital transformation in high-reliability organizations: A longitudinal study of the micro-foundations of failure," The Journal of Strategic Information Systems, vol. 32, no. 1, 2023, 101–118, Elsevier. sciencedirect.com

[3] I. Ryan, U. Roedig, and K. J. Stol, "Unhelpful assumptions in software security research," in Proc. ACM Conference on Computer and Communications Security, 2023, pp. [page range]. acm.org

[4] K. Rajwar, K. Deep, and S. Das, "An exhaustive review of the metaheuristic algorithms for search and optimization: Taxonomy, applications, and open challenges," Artificial Intelligence Review, 2023. springer.com

[5] Y. Zheng, Z. Li, X. Xu, and Q. Zhao, "Dynamic defenses in cyber security: Techniques, methods and challenges," Digital Communications and Networks, 2022. sciencedirect.com

[6] S. K. Khan, N. Shiwakoti, P. Stasinopoulos, et al., "Cyber-attacks in the next-generation cars, mitigation techniques, anticipated readiness and future directions," Accident Analysis & Prevention, 2020, Elsevier. researchgate.net

[7] R. Derbyshire, B. Green, and D. Hutchison, ""Talking a different Language": Anticipating adversary attack cost for cyber risk assessment," Computers & Security, 2021. sciencedirect.com

[8] I. F. De Arroyabe, C. F. A. Arranz, and M. F. Arroyabe, "Cybersecurity capabilities and cyber-attacks as drivers of investment in cybersecurity systems: A UK survey for 2018 and 2019," Computers & Security, vol. 124, p. 102954, 2023, Elsevier. sciencedirect.com

[9] C. Xenofontos, I. Zografopoulos et al., "Consumer, commercial, and industrial IoT (in) security: Attack taxonomy and case studies," vol. 9, no 1, pp. 199–221, IEEE Internet of Things Journal, 2021. [PDF]

[10] S. Kuipers, and M. Schonheit, "Data breaches and effective crisis communication: A comparative analysis of corporate reputational crises," Corporate Reputation Review, 2022. academia.edu

[11] K. Rindell, J. Ruohonen, J. Holvitie, S. Hyrynsalmi, et al., "Security in agile software development: A practitioner survey," Information and Software Technology, vol. 130, 2021, Elsevier. sciencedirect.com
[12] Y. Valdés-Rodríguez, J. Hochstetter-Diez, et al., "Analysis of strategies for the integration of security practices in agile software development: A sustainable SME approach," Agile Software Development: A Sustainable SME Approach. IEEE Access, 2024. ieee.org
[13] R. M. Van Wessel, P. Kroon, et al., "Scaling agile company-wide: The organizational challenge of combining agile-scaling frameworks and enterprise architecture in service companies," IEEE Transactions on Engineering Management, 2021. eur.nl
[14] M. Nakayama, E. Hustad, and N. Sutcliffe, "Organic transformation of ERP documentation practices: Moving from archival records to dialogue-based, agile throwaway documents," International Journal of Information Management, vol. 74, p. 102717, 2024, Elsevier. sciencedirect.com
[15] M. C. Annosi, E. Mattarelli, E. Micelotta, and A. Martini, "Logics' shift and depletion of innovation: A multi-level study of agile use in a multinational Telco company," Management and Organization, vol. 2022, 2022, Elsevier. sciencedirect.com
[16] N. Bussmann, P. Giudici, D. Marinelli, et al., "Explainable machine learning in credit risk management," Computational Economics, vol. 57, no 1, pp. 203–216, 2021, Springer. springer.com
[17] E. Lamine, R. Thabet, A. Sienou, D. Bork, and F. Fontanili, "BPRIM: An integrated framework for business process management and risk management," Computers in Industry, vol. 121, 2020, Elsevier. sciencedirect.com
[18] I. D. Raji, A. Smart, R. N. White, M. Mitchell, et al., "Closing the AI accountability gap: defining an end-to-end framework for internal algorithmic auditing," in Proc. 2020 Conference on Fairness, Accountability, and Transparency, 2020, pp. 33–44. acm.org
[19] D. J. Amon, S. Gollner, T. Morato, C. R. Smith, C. Chen, et al., "Assessment of scientific gaps related to the effective environmental management of deep-seabed mining," Marine Policy, vol. 138, p. 105006, 2022, Elsevier. sciencedirect.com
[20] E. Hoxha, H. R. Vignisdottir, D. M. Barbieri, F. Wang et al., "Life cycle assessment of roads: Exploring research trends and harmonization challenges," Science of the Total Environment, vol. 759, 2021, Elsevier. sciencedirect.com
[21] A. Z. Reyes, "Study on dark matter-argon interactions with effective field theories and non-thermal halo components using the DEAP-3600 experiment," 2022. unam.mx
[22] R. A. Khan, M. F. Abrar, S. Baseer, M. F. Majeed, M. Usman, et al., "Practices of motivators in adopting agile software development at large scale development team from management perspective," Electronics, vol. 10, 2021. [Online]. Available: mdpi.com
[23] L. Gren, and P. Ralph, "What makes effective leadership in agile software development teams?" in Proceedings of the 44th international conference on software engineering, pp. 2402–2414, 2022. acm.org
[24] P. Zainal, D. Razali, and Z. Mansor, "Team formation for agile software development: A review," International Journal on Advanced Science Engineering and Information Technology, 2020. semanticscholar.org
[25] H. Edison, X. Wang, and K. Conboy, "Comparing methods for large-scale agile software development: A systematic literature review," in IEEE Transactions on Software Engineering, vol. 48, no 8, pp. 2709–2731, 2021. ieee.org
[26] A. K. Bintari, and T. A. Taufik, "Analysis of business processes issuing medical equipment calibration services certificates at PT X," researchgate.net
[27] A. Limanjaya, "Balanced scorecard design at PT X construction services company," International Journal of Review Management Business and Entrepreneurship, vol. 3, no 1, pp. 87–98, 2023. uc.ac.id

[28] R. Kasauli, E. Knauss, J. Horkoff, G. Liebel et al., "Requirements engineering challenges and practices in large-scale agile system development," Journal of Systems and Software, vol. 171, 2021, Elsevier. sciencedirect.com
[29] S. Jaskó, A. Skrop, T. Holczinger, T. Chován, et al., "Development of manufacturing execution systems in accordance with Industry 4.0 requirements: A review of standard-and ontology-based methodologies and tools," Computers in Industry, vol. 123, p. 103300, 2020, Elsevier. sciencedirect.com
[30] A. Aloseel, H. He, C. Shaw, and M. A. Khan, "Analytical review of cybersecurity for embedded systems," IEEE Access, 2020. ieee.org
[31] N. K. Gopalakrishna, D. Anandayuvaraj, A. Detti, et al., ""If security is required" engineering and security practices for machine learning based IoT devices," in Proceedings of the 4th International Workshop on Software Engineering Research and Practice for the IoT, pp. 1–8, 2022. dl.acm.org. acm.org
[32] E. P. Enoiu, D. Truscan, A. Sadovykh, et al., "VeriDevOps software methodology: Security verification and validation for DevOps practices," in Dependability, Reliability and Security, 2023, pp. 1–6. dl.acm.org. acm.org
[33] I. Signoretti, L. Salerno, S. Marczak, and R. Bastos, "Combining user-centered design and lean startup with agile software development: A case study of two agile teams," in Agile Processes in Software Engineering, Springer, 2020. springer.com
[34] R. Reunamäki, and C. F. Fey, "Remote agile: Problems, solutions, and pitfalls to avoid," Business Horizons, 2023. sciencedirect.com
[35] C. Weir, S. Migues, M. Ware, and L. Williams, "Infiltrating security into development: exploring the world's largest software security study," in Proceedings of the 29th ACM Joint Meeting on European Software Engineering Conference and Symposium on the Foundations of Software Engineering, pp. 1326–1336, 2021. acm.org
[36] S. Nazir, B. Price, N. C. Surendra, and K. Kopp, "Adapting agile development practices for hyper-agile environments: lessons learned from a COVID-19 emergency response research project," Information Technology and Management, vol. 23, no 3, pp. 193–211, 2022, Springer. springer.com
[37] A. K. Tyagi, and N. Sreenath, "Cyber physical systems: Analyses, challenges and possible solutions," Internet of Things and Cyber-Physical Systems, 2021. sciencedirect.com
[38] J. P. A. Yaacoub, H. N. Noura, and O. Salman, "Ethical hacking for IoT: Security issues, challenges, solutions and recommendations," Internet of Things and Cyber-Physical Systems, Elsevier, 2023. sciencedirect.com
[39] K. Peng, M. Li, H. Huang, C. Wang, and S. Wan, "Security challenges and opportunities for smart contracts in Internet of Things: A survey," IEEE Internet of Things Journal, 2021. google.com
[40] T. Rangnau, R. Buijtenen, F. Fransen, et al., "Continuous security testing: A case study on integrating dynamic security testing tools in CI/CD pipelines," in 2020 IEEE 24th International Conference on Engineering of Complex Computer Systems (ICECCS), 2020. rug.nl
[41] J. Tang, R. Li, K. Wang, X. Gu, and Z. Xu, "A novel hybrid method to analyze security vulnerabilities in android applications," Tsinghua Science and Technology, 2020. ieee.org
[42] J. A. Morales, T. P. Scanlon, A. Volkmann, et al., "Security impacts of sub-optimal DevSecOps implementations in a highly regulated environment," in Proceedings of the 15th International Conference on Availability, Reliability and Security, pp. 1–8, 2020. dl.acm.org. acm.org
[43] M. Labouardy, "Pipeline as code: Continuous delivery with Jenkins, Kubernetes, and terraform," 2021. [HTML]
[44] M. Vielberth, F. Böhm, I. Fichtinger, and G. Pernul, "Security operations center: A systematic study and open challenges," IEEE Access, 2020. ieee.org

[45] E. Schiller, A. Aidoo, J. Fuhrer, J. Stahl, and M. Ziörjen, "Landscape of IoT security," Computer Science Review, vol. 42, 2022, Elsevier. sciencedirect.com
[46] A. Shukla, B. Katt, L. O. Nweke, P. K. Yeng, et al., "System security assurance: A systematic literature review," Computer Science Review, vol. 45, p. 100496, 2022, Elsevier. sciencedirect.com
[47] S. Al-Saqqa, S. Sawalha, et al., "Agile software development: Methodologies and trends," International Journal of Interactive Mobile Technologies, vol. 14, no 11, 2020. semanticscholar.org
[48] S. Moyo, and E. Mnkandla, "A novel lightweight solo software development methodology with optimum security practices," IEEE Access, 2020. ieee.org
[49] T. Srivatanakul, and F. Annansingh, "Incorporating active learning activities to the design and development of an undergraduate software and web security course," Journal of Computers in Education, 2022. springer.com
[50] Y. M. Tashtoush, D. A. Darweesh, G. Husari, et al., "Agile approaches for cybersecurity systems, IoT and intelligent transportation," in IEEE Access, vol. 10, pp. 1360–1375, 2021. ieee.org
[51] R. Kumar, and R. Goyal, "Modeling continuous security: A conceptual model for automated DevSecOps using open-source software over cloud (ADOC)," Computers & Security, vol. 97, p. 101967, 2020. [HTML]
[52] C. S. De Silva1, and M. T. M. Riyas, "The dummy human: Identity theft behavior patterns in a digital landscape and mitigation practices: A review," researchgate.net
[53] S. W. A. Hamdani, H. Abbas, A. R. Janjua, et al., "Cybersecurity standards in the context of operating system: Practical aspects, analysis, and comparisons," ACM Computing Surveys, 2021. dl.acm.org. researchgate.net
[54] F. A. Shaikh, and M. Siponen, "Information security risk assessments following cybersecurity breaches: The mediating role of top management attention to cybersecurity," Computers & Security, 2023. sciencedirect.com
[55] J. Haislip, J. H. Lim, and R. Pinsker, "The impact of executives' IT expertise on reported data security breaches," Information Systems Research, 2021. acm.org
[56] K. Oosthoek, and C. Doerr, "From hodl to heist: Analysis of cyber security threats to bitcoin exchanges," 2020 IEEE International Conference on Blockchain and Cryptocurrency, pp. 1–9, 2020. cyber-threat-intelligence.com
[57] A. AlQadheeb, S. Bhattacharyya, and S. Perl, "Enhancing cybersecurity by generating user-specific security policy through the formal modeling of user behavior," Array, 2022. sciencedirect.com
[58] A. H. Washo, "An interdisciplinary view of social engineering: A call to action for research," Computers in Human Behavior Reports, 2021. sciencedirect.com
[59] A. K. Kähkönen, P. Evangelista, J. Hallikas, et al., "COVID-19 as a trigger for dynamic capability development and supply chain resilience improvement," International Journal of Production Research, vol. 61, no 8, pp. 2696–2715, 2023, Taylor & Francis. tandfonline.com
[60] L. Chazette, and K. Schneider, "Explainability as a non-functional requirement: Challenges and recommendations," Requirements Engineering, 2020. springer.com
[61] E. Ismagilova, L. Hughes, N. P. Rana, et al., "Security, privacy and risks within smart cities: Literature review and development of a smart city interaction framework," Information Systems Frontiers, vol. 24, no. 2, pp. 543–576, 2022. springer.com
[62] Y. K. Dwivedi, L. Hughes, A. M. Baabdullah, et al., "Metaverse beyond the hype: Multidisciplinary perspectives on emerging challenges, opportunities, and agenda for research, practice and policy," International Journal of Information Management, vol. 62, 2022, Elsevier. sciencedirect.com

Chapter 5

DevSecOps for agile web application security

Yassine Maleh

5.1 INTRODUCTION

The waterfall software development method involves a linear flow beginning with system and software requirements and moving through design, implementation, testing, and finally maintenance leading to the application phase. Security was never an integral part of the process and always ended up being tacked on at the end. This led to the birth of the (broken) SDL (Secure Development Lifecycle) with the idea that rigorous security testing throughout the development process would lead to more secure software. The same top-down approach was taken, and when businesses began transitioning to Agile, they simply tried to use the SDL just doing security sprints stacked on top of the development sprints. This translated to doing the same iteration of security tasks on the entire application each time which doesn't match the Agile model of incremental change. This led to confusion and inconsistencies between what the security and development teams were doing and what the expectations were for each phase of the project. The end result was still ambiguous and security was still being treated as an afterthought (Díaz et al., 2019).

Rapid software development techniques, such as Agile, are successful at the expeditious delivery of applications to keep pace with the ever-changing business requirements. The traditional waterfall software development method has been unable to meet these needs. Agile developers need a development approach that allows security to "keep up" with the application changes and to build security in from the start. This is where DevSecOps comes in (Prates et al., 2019). "DevSecOps is the philosophy of integrating security practices within the DevOps process" (Ebert et al., 2016). Figure 5.1 shows DevOps lifecycle.

Traditional security practices often involved security being treated as an afterthought, but in a DevSecOps approach everyone is responsible for security and is enabled to move at a high velocity .... A DevSecOps approach fits right in with the Agile model where incremental change is welcomed. This chapter will detail the implementation of a DevSecOps approach on an Agile web application through a use case.

The chapter is organized as follows: In the next section, we delve into Understanding DevSecOps, where we outline the core principles and definitions

DOI: 10.1201/9781003478676-5

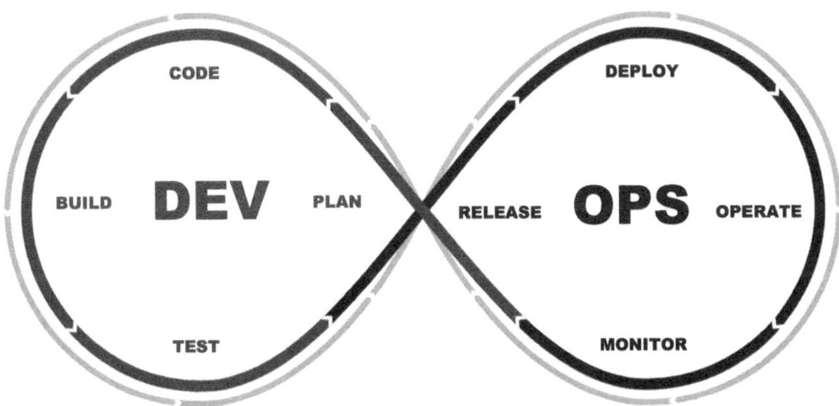

Figure 5.1 DevOps lifecycle.

essential to this approach. This is followed by a discussion on the Importance of Web Application Security, emphasizing the necessity of robust security practices. Subsequently, we explore the DevSecOps Lifecycle, detailing how security integrates into each stage of software development. The chapter then examines Implementing DevSecOps in Agile Environments, showcasing practical strategies for Agile integration. We also discuss Security Testing Techniques for Web Applications, covering various methodologies to enhance security. A practical Use Case is presented to demonstrate these principles in action, leading to the Conclusion where the findings and their implications are synthesized.

5.2 UNDERSTANDING DevSecOps

The list of benefits behind DevSecOps is something that security and risk management teams can relate to and effectively want to ensure that security practices do not stand in the way of project timelines and outcomes (Hsu, 2018). This is commonly the main issue why security activities find it difficult to be integrated within Agile projects and can cause various issues for development teams when subjected to traditional security assessment and testing activities toward the end of a project (Sahid et al., 2020b). One of the key reasons for this is that security is often viewed as a set of restrictions or as "blocking" activity. When security does find itself a requirement for a project, it can often result in costly rework of code to fix various security vulnerabilities. DevSecOps tries to push security as a core requirement by implementing automatic security testing throughout the development of a project in the form of continuous integration/deployment (CI/CD) and delivering immediate feedback to developers on security-related issues. This will result in fewer bugs and security vulnerabilities being introduced and, if successful, would

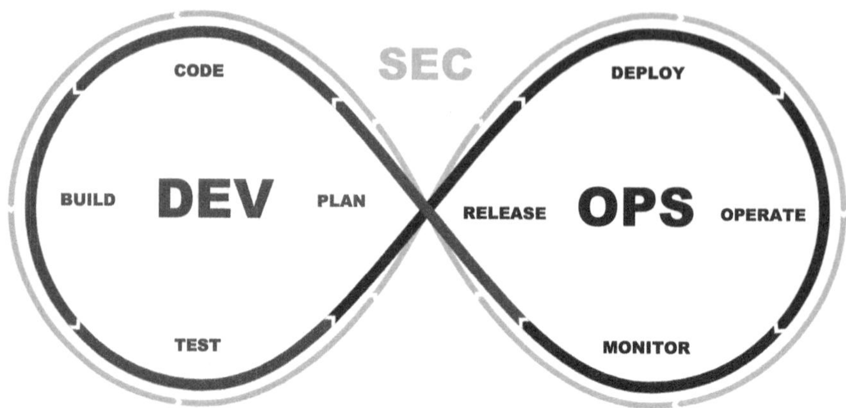

Figure 5.2 DevSecOps lifecycle.

prevent the common change and rework of security-related requirements due to time and resource constraints (Figure 5.2).

Understanding the concept of DevSecOps involves an appreciation of its basic definition, the principles that support it, and how it is integrated with both Agile methodologies and traditional software development and system operations. To describe the definition of DevSecOps is, in essence, the application of a security layer over the top of a solid Agile or DevOps practice. It can also be seen as a cultural change of merging roles to break down the traditional silos that have typically existed between development, QA, and security teams. The latter view is supported by a more detailed description of DevSecOps principles expressed by John Willis, who describes it as a set of practices that automates and integrates security at every phase of the software development lifecycle (SDLC). Here, it is important to understand the differences between doing security in a DevSecOps culture and that of traditional security assurance.

5.2.1 Definition and principles

DevSecOps was introduced to cater to the need to speed up the duration for software to reach a security operations team. It represents a shift in IT culture to instill collaboration among siloed teams: development, IT operations, and security teams. This is also characterized by a union of automation and monitoring at all steps of software construction, as well as tightly integrating security into every process lifecycle. DevSecOps also means that security becomes everyone's responsibility, and not just the security team's job. With this, it is undeniable that DevSecOps is described as putting a piece of security into DevOps and IT operations.

The word DevSecOps is made up of three words. "Dev" is usually defined as software development. "Sec" is usually short for security. "Ops" is a term that can mean IT operations, though the definition is broader and can involve healthy discussions about who bears the responsibility for new code. Combining the three, we come up with the definition that DevSecOps is a set of practices that seeks to automate and integrate the tasks between software development, IT operations, and the security teams in order to release secure software at a faster rate (Abdelkebir et al., 2019).

5.2.2 Benefits and challenges

Security, such as detailed design specs, will be done at the beginning of a phase and integrated with static analysis and dynamic analysis tools. The output will be that security tests uncover very few issues, making developers annoyed that security is finding bugs after code is locked down. The cost to fix bugs is increasing significantly (Sahid et al., 2020a).

In traditional security, the more up-front analysis, the more likely you are to get security baked in. However, the longer the phase, the more the late-stage cost goes up. The annoyance is security being told late that they have to redo something because requirements have changed. At this point, the cost to change is significantly higher because development has been completed and various costs have been incurred.

Security, in general, is not something most developers are excited about. Taking time to code, add, and test security often results in excitement and productivity decrease, as well as reduced freedom in coding. Oftentimes, security is ignored until it is absolutely needed. This can be a disaster and often causes significant delays in deployment. The longer the wait, the more risk and cost to fix security issues.

5.2.3 Integration with Agile methodologies

Integrating security into an Agile environment can be simplified by using security-focused user stories, which can be interwoven with general user stories throughout the project. This enables the security team to work closely with the development team, advising and implementing secure code as each bit of functionality is built. Security automation can greatly assist in Agile development's iterations and fast-paced nature. By using continuous testing throughout a development cycle, security defects can be found and addressed much quicker than traditional testing methods (Read et al., 2016). This is much more effective than trying to test a mass of code at the end of a development cycle and find security, as well as functionality defects. This can be achieved by having the security team build security-based test cases which can be automated alongside general QA test cases.

Agile development uses an iterative approach to project management and operations that can be very beneficial to the DevSecOps cycle. Because Agile development has the flexibility to add or change small bits of code and functionality throughout the SDLC as shown in Figure 5.3, it can add security bit by bit alongside the new features.

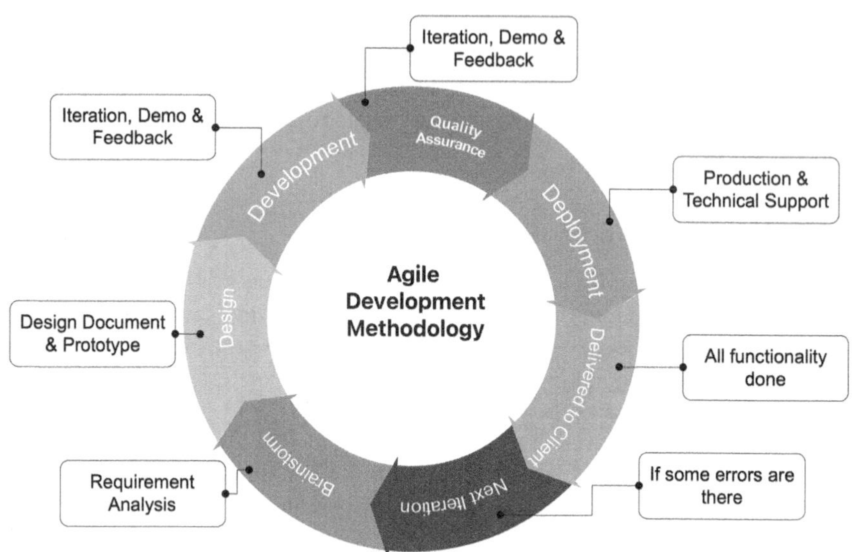

Figure 5.3 SDLC methodology.

This is especially important with security, as bolting on security at the end of a development cycle is never as effective as building it in from the start. Agile methodologies are driven by iterations and continuous improvement, something that is very much in line with inspecting and adapting in the DevSecOps cycle (Sahid et al., n.d.). Agility in development and having security as part of that agility is paramount in today's fast-paced world.

5.3 IMPORTANCE OF WEB APPLICATION SECURITY

A key driver to the widespread interest in application security is the increasing number of threats and attacks on application-level vulnerabilities that result from a poor focus on application and information security by many organizations (Sadqi & Maleh, 2022). A number of high-profile hacking attacks have also brought increased attention to application security. Invariably, data usage over the internet and web services will continue to expand, and the increasing frequency of stories about identity theft and security breaches will only fuel the demand for more security. This is why application security should be a key concern. Going forward, the most efficient and effective approaches to tighten security without slowing application delivery will be the ones that win the most support. That is why it is crucial to have technologies in place that support the principles of secure application development, and that application security is at the forefront of all technology decisions. DevOps and the recent emergence of "DevSecOps" are a move in that direction and have the potential to make application security a far more prevalent concern in the future than it is today.

5.3.1 Risks and vulnerabilities

The ultimate goal of an attacker would be system-level access and the use of the system as a resource to perform tasks for an extended period of time. This level of impact is actually quite rare given the knowledge of today's attackers and tools. However, past instances of this type of attack have led to complete system failure and the end of the affected service (Yassine et al., 2017).

The impact of a security breach on a web application can vary from not much at all to complete system failure. The true damage is determined by the value of the information or the service to the attackers, and the level of harm they wish to inflict. Information is a valuable asset. If a system stores information about something that has monetary value, then it is at risk of that information being stolen. This is the same for personal information and identity information. People and systems providing web services can be a target if the service is valuable, as it may be defaced or disrupted with malware.

Web application attacks have increased in recent years. This is due to the sheer number of potential vulnerabilities and not the awareness of security by the developers. Though security is not the only factor that has driven these attacks, others include hacktivism, organized crime, and national security. But the fact remains that there are a large number of web application vulnerabilities and a relatively low awareness of web application security.

In simple terms, a risk is a potential harmful event. Vulnerability is a way in which the system may be affected by such an event, and impact is the amount of harm caused by the event. When we speak about risks to our web applications, there are a large number of both known and unknown vulnerabilities that can be exploited to affect the system in adverse ways. Due to the number of ever-increasing web applications and the technologies to support them, web applications have a large variety of ways in which they can be implemented. This disparity in implementation can lead to a higher occurrence of vulnerabilities.

5.3.2 Impact of security breaches

Large-scale and complex web application systems are becoming increasingly integral to everyday life as more people access and utilize the web as a critical business tool. Cloud computing, social networking, and various Web 2.0 applications are fueling this trend. Unfortunately, the accompanying risks and security threats are also growing in both frequency and impact. Data breaches, defacement of publicly accessible resources, and the constant targeting of online services are just a few examples of security incidents that have a significant impact on individuals and businesses (Patel et al., 2010). Security breaches can result in expensive and sometimes irreversible damage to an organization's reputation. Within the most severe cases, a security incident can put an organization out of business. Such organizations may never fully recover the loss of intellectual property, data, or additional resources. Industries that handle customers' personal data have a legal and ethical obligation to protect that data. Net and web service security within the health industry is an example where sensitive patient records are an attractive target to malicious parties. Any data breaches in this sector can have a long-term impact on both the affected individuals and the organizations involved. It can be seen that the impact of a security incident can vary widely, but for the vast majority of cases, the implications are predominantly negative.

5.3.3 Compliance requirements

As Jamil Aslam explains, DevSecOps can automate these challenges into the software delivery lifecycle. Policy is translated into test-driven security behavior that is unit and acceptance tested alongside application code. Automation scripts can test and verify regulatory controls in live environments, and data from these testing activities can provide the needed evidence that the software is compliant or identify areas where more work is needed. This proactive approach replaces the more common "bolted on" security and post-development verification of compliance, with increased cost-effectiveness and assurance of meeting the intended compliance level (Humphreys, 2008).

Adherence to today's often complex legal mandates and industry standards is a fact of life for many companies. With each line of new or revised legislation, businesses must take action by interpreting the regulation in terms of policies and system changes, auditing for both the policy changes and their effectiveness in realizing compliance, and at all times being prepared to demonstrate their "evidence" of due diligence in the event of an incident and regulatory review.

DevSecOps helps address these compliance requirements in a lower cost and more comprehensive manner than traditional approaches to application security. By baking security and the automation of compliance controls into the software delivery process, DevSecOps reduces the total cost and effort of achieving secure software and systems and provides a real-time compliance dashboard to executives and compliance officers.

Laws such as the US Health Insurance Portability and Accountability Act (HIPAA) (HIPAA, 1999), EU Data Protection Directive, and the Australian Privacy Act require the protection of personal information through appropriate security measures. Other generic regulations such as the Sarbanes-Oxley (SOX) Act and PCI Data Security Standard (Gikas, 2010) apply to a wider range of private data in information systems, while the more recent EU General Data Protection Regulation is applicable to any organization that collects data from EU citizens. Failure to comply with these laws may result in significant fines and damaged reputation.

Businesses and organizations may be required to comply with legal and regulatory requirements that mandate the security and privacy of personal data. Non-compliance may result in severe penalties, including financial damages and legal liabilities. It is crucial for such organizations to understand the regulatory environment, assess how they are affected by it, and then act to ensure compliance.

5.4 DevSecOps LIFECYCLE

Deployment using the cloud is cost-efficient and easily changeable through the internet, but most deployment activities do not involve security: no security testing, no access control configurations, etc. DevSecOps suggests deploying small features that can be tested for security and functionality. This is to reduce a lot of rework in the future if many features were deployed at the same time and had to be removed or debugged due to failures.

DevSecOps changes the way how deployment and monitoring are viewed in agile today. In agile, we know that development and deployment should be done in parallel to reduce the timeline. Now, with cloud computing, this may seem more challenging and cause a lot of configuration and access control issues if security is not involved early.

DevSecOps would like to enhance the risk-based approach in agile by providing quantitative and qualitative data to support the risk level decision and have features in agile to redefine project or sprint decisions based on new security risks and benefits. Threat modeling will produce threat intelligence throughout the lifecycle. It is important to define security stories and tasks based on the threat modeling output for tracking and reporting purposes.

DevSecOps takes planning and design to the next level, where security conceptual ideas begin to integrate throughout the application lifecycle. By starting with a variety of risk-based approaches, including threat modeling and metrics development, the focus shifts to identifying effective ways to integrate security activities into the existing agile plan, such as incorporating user stories or security-based tests as features.

The SDLC is often categorized into six phases: analysis, design, implementation, testing, deployment, and maintenance. DevSecOps should find a way to integrate and automate security throughout the entire lifecycle in an agile way. This could be a challenge as agile promotes adaptive planning, evolutionary development, early delivery, and continual improvement, and it encourages a flexible response to change.

5.4.1 Planning and design

Once in the design stage, it is typical to spend a long time creating and revising a site's information architecture and design wireframes. With a good bidder or tendering process, work for suppliers would be limited to creating simple prototypes to try and win design contracts, rather than doing full-scale development work. It is common that the better frontend developers get the most money for work that they do the least of, but this is not our topic here.

A typical project would spend several months planning and documenting to gather and make sure that they understood the requirements correctly before moving into the design phase. They aim to try and understand all the requirements upfront to minimize requirement changes, lower the risk of project failure, and reduce costs incurred.

In larger companies, it is not unusual for new websites or extensive site updates to be planned over a year or more in advance. A waterfall project management methodology would typically have a long, drawn-out planning phase, followed by an even longer design and prototyping phase.

5.4.2 Development and testing

CD, in which every successful build is deployed to production, is widely regarded to be the final stage of automation for an agile project. High-performing organizations deploy much more frequently than low performers, with lower change failure rates and with much less time required to recover from failed changes. Although there are no statistics relating specifically to the frequency of changes to security configurations and code changes which affect security, the automation of deployment combined with shorter cycles would appear to be beneficial so long as it does not compromise the thoroughness of other security activities.

Agile development emphasizes automation and testing, and it is natural to see these ideas carried through to security testing. Continuous integration, in which code from developers is frequently integrated, tested, and checked into version control, can be countered by the deployment of automated security tests. If the CI pipeline is further automated to the extent that, on successful security tests, the code is automatically deployed into a staging environment, security regression testing can be carried out to frequently check that changes to the infrastructure and application have not undermined security in a way that is difficult to detect with periodic penetration testing.

5.4.3 Deployment and monitoring

Monitoring applications and systems in the cloud is very different from monitoring on-premises resources. Unlike on-premises resources, when you monitor applications and workloads that are running in the cloud, you no longer have direct visibility or control of the underlying infrastructure. In a cloud environment, services are ephemeral, meaning they may only exist for a few seconds, minutes, or hours. Services can fail at any time and be replaced by other identical instances. Traditional monitoring systems can't cope with situations where resources are temporary and frequently changing. On top of this, cloud-based systems are often highly distributed and consist of many moving parts. More traditional systems and applications may work with a single database and a few servers, whereas modern, cloud-based applications may use services such as queues, NoSQL databases, or microservices that each have their own autoscaling instances. The complexity of these systems makes monitoring difficult but is also a reason why effective monitoring—particularly the ability to gain insights into the relationships between different components—is crucial.

5.5 IMPLEMENTING DevSecOps IN AGILE ENVIRONMENTS

The objective is to produce higher quality secure code with lower risk to overall project cost and time.

Traditional security testing would rely on developers halting their changes for the next release and engaging a security team to perform testing, often resulting in an ineffective iterative cycle between development and security. Dynamic analysis and testing can still be effective in agile methods with a shift toward automation of tests that can be run frequently against code in development. This can ensure that agile changes do not result in security regression and aid in maintaining visibility of risk to development teams.

Static analysis and continuous monitoring can promote a more proactive security approach in comparison to traditional methods. Static analysis can be performed by tools in an offline manner on software artifacts where findings are reported back to developers as a list of issues to be fixed. Ongoing development can still continue, and developers are able to use the reports to manage the security posture of their software and integrate changes as required.

By implementing security as code practices, security policies become associated with development environments and the software itself. This can be delivered using agile automation techniques to provide a consistent and auditable security posture. During software development phases, security requirements can be managed alongside other code requirements in an agile manner. This helps to integrate security practices within agile development, as opposed to security being a post-development activity.

DevSecOps is mainly aimed to bridge the gap with agile development and is often a misaligned target. In early phases, it was difficult to align traditional security practices with fast-paced Dev practices. One of the key strengths behind agile methodologies is its ability to manage changes in requirements for software development projects. This concept is often misaligned with traditional security practices, which revolve around reducing changes to IT environments to a minimum in order to maintain a known good security posture. This often results in a conflict of interest between development teams and security teams, ultimately where security is seen as a bottleneck to agile processes.

5.5.1 Collaboration between development, security, and operations teams

Establishing collaboration between the developers, security professionals, and IT operations is a key part of implementing DevSecOps in secure software development. Involvement of the development team in security practices provides them with an understanding of security issues and how to address them. It also gives them a sense of accountability for security in their code. The goal is to equip developers with the practices and controls they need to write and test secure and resilient code. To achieve this, security tasks need to be woven into day-to-day development activities. Security APIs will be provided for developers, and security libraries, frameworks, and tools will be infused into the development environment. Training developers to code more securely is also an essential part of this practice. Another method of integrating development and security practices is to use security automation in development processes. This involves automating security tools and processes, and integrating them into the developer's environment. The goal is to give the developer real-time feedback on security-related issues, as he or she is writing the code. This allows security issues to be resolved early and often and minimizes the chance of security-related defects making their way into production. At its core, security automation is about building security controls and processes directly into the software development process. It leverages the agility of the development team, with the goal of making security a transparent and intrinsic element of development processes.

5.5.2 Continuous integration and delivery

Adopting CI/CD is an important step in securing Agile web applications because often one of the biggest pressures on security testing is the project timeline and there will always be a tradeoff between time and quality of testing. By automating the testing, and even the promotion of tested code, we are able to alleviate this time pressure and largely increase the amount of testing done on each increment of code without impeding developers' progress. At the most basic level, security tests can be integrated with existing tests, providing immediate learning of security-related bugs. More advanced security tests for web applications such as static analysis security testing (SAST) and dynamic analysis security testing (DAST) require some preparation, but once integrated can be run frequently for little cost. DAST can be run every time the application is deployed to the testing environment to quickly detect new vulnerabilities, and SAST tools can be set to watch the changes in the code and run on demand or with regularity. So long as the security tests have reasonable coverage and reliability, the result is a significant increase in security testing frequency and depth.

CI/CD model is a software development strategy. For our purposes, we will consider CI/CD to mean the automation infrastructure for our application testing and deployment, which runs whenever new code is integrated into the application or its environment. A simple example of this type of automation would be the integration of JUnit or TestNG. A more complex example is a deployment pipeline which runs a series of automated steps in different environments to fully test the application and facilitate a decision to promote the new code to production.

5.5.3 Automating security testing

SAST is the automated testing of source code and object code for security vulnerabilities. At the source code level, the code is analyzed for typical coding mistakes which can lead to vulnerabilities. Object code analysis involves the testing of an application which has already been compiled. SAST tools generally generate a large number of false positives and also require a reasonable level of security expertise to assess and mitigate found issues. However, they are still a valuable security testing tool. High false positive rates may slow down the development process as these issues may be investigated by developers and still not be valid security issues. Requirements to ensure security expertise is present in development teams are achievable using a balanced mix of training existing staff and employing new staff with the required security knowledge. This will allow for a greater understanding of security issues in development teams and will ultimately lead to a higher level of security in developed software.

With the increasing complexity of web applications and the need for quick delivery, automating security tests is an important part of any DevSecOps strategy. The goal of security automation is to execute security tests with little to no human intervention, continuously in a pre-production or production environment. Many software developers and testers struggle to competently test for security-related issues in their web applications due to a lack of personnel with the required skills. Automating security tests allows testers to pinpoint security issues early in the development process, at a point in time where fixing these problems is significantly cheaper than after deployment.

5.6 SECURITY TESTING TECHNIQUES FOR WEB APPLICATIONS

Static code analysis tools automatically search through a program's code and then look for possible security vulnerabilities, including whether the code could leak personal data, perform illegal operations, or provide an open backdoor to the software. These tools can be language-specific or general analysis tools, and they build a representation of how the code is expected to behave when running and then work out whether there are any deviations from this behavior. They provide a list of potential issues with recommendations on how to fix each problem.

In this context, security issues can be picked up and fixed as and when they are found without needing to go back a significant amount of development time to correct issues. This is in contrast to finding security issues through penetration testing, which can be time-consuming in the later stages of development to fix and lead to dissent from the development team who are against making changes to a now complete system.

Static code analysis can be effectively placed in the middle of the development process and early on in the requirements phase. This is an automated method of code analysis that looks for security vulnerabilities in code in a non-running state. This has the effect of finding security issues early in the SDLC. This is important as it is an unfortunate feature of many projects that software security is left as an afterthought right up until the maintenance phase or not considered at all and instead is focused on functionality.

5.6.1 Static code analysis

Static code analysis is the practice of testing source code for quality, reliability, and security. This test is usually run in the earlier stages of software development and often used in Agile development approaches. DevSecOps looks at injecting SAST tools directly in the development pipeline (i.e. left shifting) to find and fix security issues early in the development process. The idea is that by using SAST tools, the code is constantly being checked for security vulnerabilities. SAST tools work by scanning the source code, byte code, and binaries of an application looking for conditions that typically result in vulnerabilities. They do this by using pattern matching to syntax tree analysis. Issues that are generally categorized as false positives are then reported to the development team. Since these issues are being reported directly to the development team, they can be fixed as soon as they are written. This approach is far more efficient and cost-effective. The further into the development cycle a security issue is found, the more expensive it is to fix. SAST tools primarily look at finding security vulnerabilities, whereas traditional SCA tools also check code for its quality and reliability. The idea of migrating away from traditional SCA tools and using SAST tools is a positive step forward for improving security in software as it is far more efficient and produces better results in finding and fixing security issues.

5.6.2 Dynamic application security testing (DAST)

The DAST is the technique to identify the vulnerabilities in the web application while the application is running. DAST is also known as black box testing. It is an effective means of finding vulnerabilities in a web application. It is performed during the testing phase of security testing. DAST is an easy process to implement but needs to be executed constantly because the dynamic aspects of modern web applications can cause code changes to break the results of previously successful tests. Changes to test data can also create challenges in comparing historical results. Dynamic testing includes both manual and automated testing, and we can execute it using internal or external methods. Internal methods have to create the test data, while external methods are using the internet and information shared by companies to get the opportunity to learn about the site and devise test cases.

5.6.3 Interactive application security testing (IAST)

IAST involves analyzing an application in a "non-running" state (Li, 2020). It is highly useful at identifying security vulnerabilities in an application's source code; however, it does not provide any information about how the application behaves and executes at runtime. Static testing analyzes the application in a "non-running" state and is often very cost-effective and can be integrated well with quality assurance and application development lifecycle. SAST is suited to Agile development due to the ability to quickly scan the source code and identify security vulnerabilities before the iteration is complete. Static code analysis can affect the effectiveness at which DAST can identify vulnerabilities and efficiency of false positive removal by reducing the number of security vulnerabilities present in the source code and identifying security vulnerabilities directly after the code has been written. Static code analysis can affect the effectiveness at which DAST can identify vulnerabilities by reducing the number of security vulnerabilities present in the source code. Using SAST after an iteration is complete to identify vulnerabilities is valuable; however, it may hinder progress as developers must backtrack to fix security vulnerabilities.

IAST is a term that describes a comprehensive testing method that combines DAST and SAST testing internal to the application. In reality, IAST is the next logical step after utilizing DAST and SAST because it takes the best features of both testing methods to create a more in-depth understanding of the application. This method is accomplished through the use of a web application or application security testing tool that is able to monitor and identify security vulnerabilities in real time as the application is running.

5.7 USE CASE

5.7.1 DevSecOps pipeline overview

The DevSecOps pipeline is an automated, continuous process integrating software development, operations, and security throughout the application lifecycle (Figure 5.4).

Here is an overview of the key steps of the pipeline used in this project:

- Source Code Management: The process starts with source code management using platforms such as Git, where developers collaboratively contribute to the code. Practices such as version control are applied to track and manage changes.
- CI: For each change in the source code, CI is launched via Jenkins. Automated testing, including unit testing and SAST, is performed to ensure code quality and to identify potential vulnerabilities.

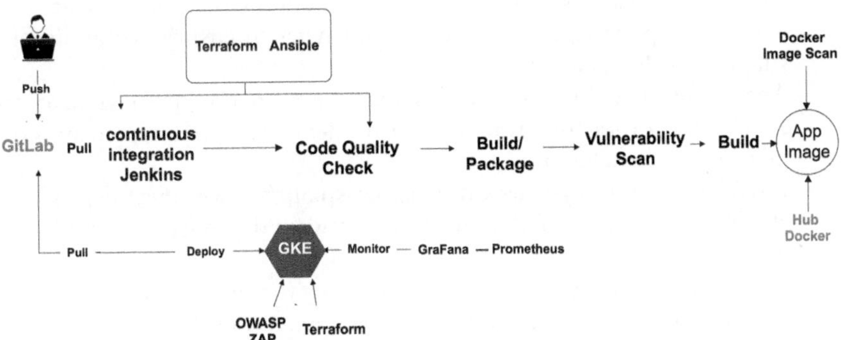

Figure 5.4 The proposed DevSecOps pipeline.

- Infrastructure Provisioning: Infrastructure is provisioned and managed as code using Terraform to provide consistent and automated management of resources in the AWS cloud environment.
- Container Orchestration: Kubernetes is used for container orchestration, providing efficient and scalable management of contained workloads throughout deployment.
- DAST: Once applications are deployed, the DAST is performed using specific tools to assess the security of real-world applications.
- Continuous Delivery (CD): After the testing phase, CD automatically deploys the validated applications to the production environment.
- Monitoring and Tracking: Once in production, monitoring mechanisms are put in place to detect any anomalies and collect data on application performance and security.

This DevSecOps pipeline represents a holistic approach to software development, where security is integrated at every stage, enabling fast and secure application delivery.

5.7.2 Tools

At the heart of the DevSecOps pipeline is a range of carefully selected tools to ensure the development process's quality, speed, and security. Each of these tools fulfills a specific role, contributing crucially to the creation of a solid and secure development ecosystem:

- SonarQube (https://www.sonarsource.com/open-source-editions/sonarqube-community-edition): A static source code analysis tool that identifies code quality issues, security vulnerabilities, and potential errors.
- Checkstyle (https://checkstyle.sourceforge.io): An automatic Java source code checker tool that applies defined style rules to ensure code consistency and quality.
- Docker (https://www.docker.com/): A containerization platform for isolating applications and their dependencies to ensure portability and simplified management.
- Apache Maven (https://maven.apache.org/): Project management and construction tool for Java, facilitating dependency management, compilation, and packaging.
- OWASP Dependency-Check (https://owasp.org/www-project-dependency-check/): A security tool that identifies software dependencies with known vulnerabilities.
- Jenkins (https://www.jenkins.io/): A CI server that automates the process of building, testing, and deploying applications.
- Trivy (https://trivy.dev/): A vulnerability scanner for containers is used to identify security vulnerabilities in Docker images.

- OWASP ZAP (https://www.zaproxy.org/): An automated security testing tool to find vulnerabilities in web applications (Jakobsson & Häggström, 2022).
- Argo CD (https://argo-cd.readthedocs.io/en/stable/): GitOps operations tool for CD and configuration management of Kubernetes applications.
- Grafana and Prometheus (https://grafana.com/docs/grafana/latest/getting-started/get-started-grafana-prometheus/): Grafana is a data visualization platform, while Prometheus is a monitoring and alerting system.

5.7.3 Infrastructure

The underlying infrastructure, deployed and managed with Terraform, is based on cloud architecture. This architecture uses resources such as virtual instances, container services, managed databases, and storage solutions, ensuring consistent flexibility and reproducibility:

- GitLab (https://about.gitlab.com/): An application lifecycle management platform with source code management, CI, and CD.
- Ansible (https://www.ansible.com/): Automation of infrastructure configuration and management, enabling the deployment and management of complex systems.
- Terraform (https://www.terraform.io/): An infrastructure-as-code (IaC) tool for defining and provisioning cloud infrastructures in a declarative manner.
- Google Cloud Platform (https://cloud.google.com/): A set of cloud services provided by Google that offers storage solutions, computing, databases, analytics, and more.
- Google Kubernetes Engine (GKE) (https://cloud.google.com/kubernetes-engine): A Kubernetes-based container management service for deploying and managing containerized applications on Google Cloud.

5.7.4 Establishing the DevSecOps Environment

5.7.4.1 Installing Terraform

To install Terraform, HashiCorp distributes Terraform as a binary package. You can also install it using APT package manager. Ensure that your system is current and that you have installed the GnuPG, software-properties-common, and curl packages. You will use these packages to verify HashiCorp's GPG signature and install HashiCorp's Debian package repository.

5.7.4.1.1 Writing Terraform configuration files

Create the Terraform configuration files to define the necessary infrastructure for Jenkins and SonarQube. Files in HCL (HashiCorp Configuration

Name	Last commit	Last update
..		
modules	Revert "Delete main.tf"	1 week ago
.gitkeep	Add new directory	2 weeks ago
main.tf	Update main.tf	2 weeks ago
provider.tf	Update provider.tf	2 weeks ago
terraform.tfvars	Revert "Delete terraform.tfvars"	1 week ago
variables.tf	Update variables.tf	2 weeks ago

Figure 5.5 Configuration Terraform.

Language) format are used to write the required resources, such as VMs and K8s cluster (Figure 5.5).

5.7.5 Initializing Terraform and applying changes

Before you unload the infrastructure, run the following commands from the repository containing your Terraform files, as shown in Figure 5.6.

This will initialize Terraform and download the necessary providers. Before actually deploying the infrastructure, it is recommended that a Terraform plan be managed. This step allows you to preview the changes proposed by your Terraform configuration (Figure 5.7).

Finally, apply the changes (Figure 5.8).

5.7.5.1 *Infrastructure configuration*

Once the infrastructure is created using Terraform, we will configure the Jenkins and SonarQube servers using Ansible.

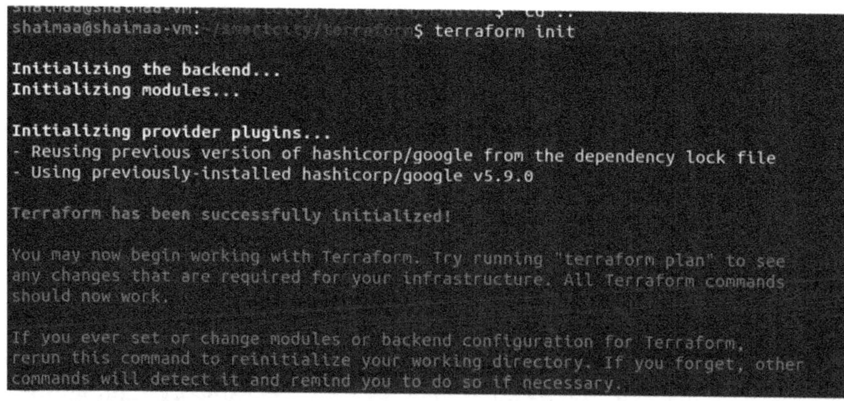

Figure 5.6 Terraform init.

```
# module.gke_cluster.google_container_cluster.gke_cluster will be created
+ resource "google_container_cluster" "gke_cluster" {
    + cluster_ipv4_cidr              = (known after apply)
    + datapath_provider              = (known after apply)
    + default_max_pods_per_node      = (known after apply)
    + deletion_protection            = true
    + enable_intranode_visibility    = (known after apply)
    + enable_kubernetes_alpha        = false
    + enable_l4_ilb_subsetting       = false
    + enable_legacy_abac             = false
    + enable_shielded_nodes          = true
    + endpoint                       = (known after apply)
    + id                             = (known after apply)
    + initial_node_count             = 2
    + label_fingerprint              = (known after apply)
    + location                       = "us-central1-f"
    + logging_service                = (known after apply)
    + master_version                 = (known after apply)
    + monitoring_service             = (known after apply)
    + name                           = "smartcity-cluster"
    + network                        = "default"
    + networking_mode                = "VPC_NATIVE"
    + node_locations                 = (known after apply)
    + node_version                   = (known after apply)
    + operation                      = (known after apply)
    + private_ipv6_google_access     = (known after apply)
    + project                        = (known after apply)
    + remove_default_node_pool       = true
    + self_link                      = (known after apply)
    + services_ipv4_cidr             = (known after apply)
    + subnetwork                     = (known after apply)
    + tpu_ipv4_cidr_block            = (known after apply)
```

Figure 5.7 Terraform plan.

```
Plan: 2 to add, 0 to change, 0 to destroy.

Do you want to perform these actions?
  Terraform will perform the actions described above.
  Only 'yes' will be accepted to approve.

  Enter a value: yes

module.gke_cluster.google_container_cluster.gke_cluster: Creating...
module.gke_cluster.google_container_cluster.gke_cluster: Still creating... [10s elapsed]
module.gke_cluster.google_container_cluster.gke_cluster: Still creating... [20s elapsed]
module.gke_cluster.google_container_cluster.gke_cluster: Still creating... [30s elapsed]
module.gke_cluster.google_container_cluster.gke_cluster: Still creating... [40s elapsed]
module.gke_cluster.google_container_cluster.gke_cluster: Still creating... [50s elapsed]
module.gke_cluster.google_container_cluster.gke_cluster: Still creating... [1m0s elapsed]
module.gke_cluster.google_container_cluster.gke_cluster: Still creating... [1m10s elapsed]
module.gke_cluster.google_container_cluster.gke_cluster: Still creating... [1m20s elapsed]
module.gke_cluster.google_container_cluster.gke_cluster: Still creating... [1m30s elapsed]
module.gke_cluster.google_container_cluster.gke_cluster: Still creating... [1m40s elapsed]
module.gke_cluster.google_container_cluster.gke_cluster: Still creating... [1m50s elapsed]
module.gke_cluster.google_container_cluster.gke_cluster: Still creating... [2m0s elapsed]
module.gke_cluster.google_container_cluster.gke_cluster: Still creating... [2m10s elapsed]
module.gke_cluster.google_container_cluster.gke_cluster: Still creating... [2m20s elapsed]
module.gke_cluster.google_container_cluster.gke_cluster: Still creating... [2m30s elapsed]
module.gke_cluster.google_container_cluster.gke_cluster: Still creating... [2m40s elapsed]
module.gke_cluster.google_container_cluster.gke_cluster: Still creating... [2m50s elapsed]
module.gke_cluster.google_container_cluster.gke_cluster: Still creating... [3m0s elapsed]
module.gke_cluster.google_container_cluster.gke_cluster: Still creating... [3m10s elapsed]
module.gke_cluster.google_container_cluster.gke_cluster: Still creating... [3m20s elapsed]
module.gke_cluster.google_container_cluster.gke_cluster: Still creating... [3m30s elapsed]
module.gke_cluster.google_container_cluster.gke_cluster: Still creating... [3m40s elapsed]
module.gke_cluster.google_container_cluster.gke_cluster: Still creating... [3m50s elapsed]
```

Figure 5.8 Terraform apply.

5.7.5.2 *Installing Ansible*

To get started with Ansible and manage your server, you must install the Ansible software on the machine that will serve as the Ansible control node.

5.7.5.3 *Writing Ansible playbooks*

Create Ansible playbooks to write specific configuration tasks for Jenkins, SonarQube, Docker, and Trivy. The playbooks are written in YAML format, and the steps to follow are written down to configure each component.

5.7.5.4 *Executing the configuration*

After writing the Ansible playbooks, run them to apply the configuration on the corresponding server (Figures 5.9–5.11).

```
shaimaa@shaimaa-vm:$ ansible-playbook -i inventory trivy.yml

PLAY [all] ************************************************************

TASK [Gathering Facts] ************************************************
ok: [34.41.83.134]

TASK [Download Trivy Debian package] **********************************
changed: [34.41.83.134]

TASK [Install Trivy] **************************************************
changed: [34.41.83.134]

PLAY RECAP ************************************************************
34.41.83.134          : ok=3    changed=2    unreachable=0    failed=0    skipped=0    rescued=0    ignored=0
```

Figure 5.9 Installing Trivy.

```
shaimaa@shaimaa-vm:$ ansible-playbook -i inventory docker.yml

PLAY [Docker Installation] ********************************************

TASK [Gathering Facts] ************************************************
ok: [34.41.83.134]

TASK [update] *********************************************************
changed: [34.41.83.134]

TASK [install ca-certificates curl gnupg] *****************************
ok: [34.41.83.134]

TASK [Create directory /etc/apt/keyrings with permissions 0755] *******
changed: [34.41.83.134]

TASK [Download Docker GPG key] ****************************************
ok: [34.41.83.134]

TASK [Import Docker GPG key] ******************************************
changed: [34.41.83.134]

TASK [Set permissions for Docker GPG key] *****************************
ok: [34.41.83.134]

TASK [Add Docker repository to sources.list.d] ************************
changed: [34.41.83.134]

TASK [update] *********************************************************
changed: [34.41.83.134]

TASK [install ca-certificates curl gnupg] *****************************
changed: [34.41.83.134]

PLAY RECAP ************************************************************
34.41.83.134          : ok=10   changed=6    unreachable=0    failed=0    skipped=0    rescued=0    ignored=0
```

Figure 5.10 Installing Docker.

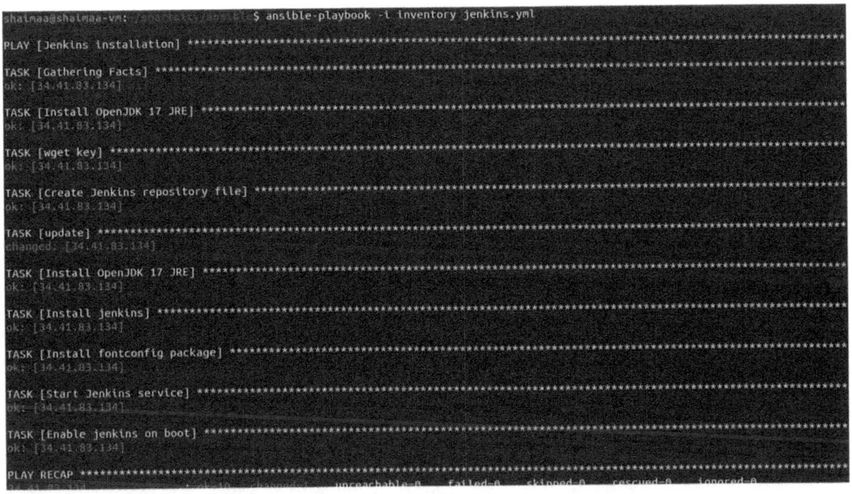

```
shaimaa@shaimaa-vm:~/smartcity/ansible$ ansible-playbook -i inventory jenkins.yml

PLAY [Jenkins installation] ****************************************************

TASK [Gathering Facts] *********************************************************
ok: [34.41.83.134]

TASK [Install OpenJDK 17 JRE] **************************************************
ok: [34.41.83.134]

TASK [wget key] ****************************************************************
ok: [34.41.83.134]

TASK [Create Jenkins repository file] ******************************************
ok: [34.41.83.134]

TASK [update] ******************************************************************
changed: [34.41.83.134]

TASK [Install OpenJDK 17 JRE] **************************************************
ok: [34.41.83.134]

TASK [Install jenkins] *********************************************************
ok: [34.41.83.134]

TASK [Install fontconfig package] **********************************************
ok: [34.41.83.134]

TASK [Start Jenkins service] ***************************************************
ok: [34.41.83.134]

TASK [Enable jenkins on boot] **************************************************
ok: [34.41.83.134]

PLAY RECAP *********************************************************************
```

Figure 5.11 Installing Jenkins.

5.7.6 Testing and building the application

5.7.6.1 Installing Maven

Install Maven on the Jenkins server to facilitate the application build process (Figure 5.12).

5.7.6.2 Setting up SAST

In this section, we'll set up SonarQube, integrate Checkstyle, and use OWASP Dependency-Check to enhance the security of our application.

5.7.7 Configuring SonarQube

5.7.7.1 Plugin installation

Before you start integrating SonarQube with Jenkins, make sure you install the appropriate SonarQube plugin (Figure 5.13).

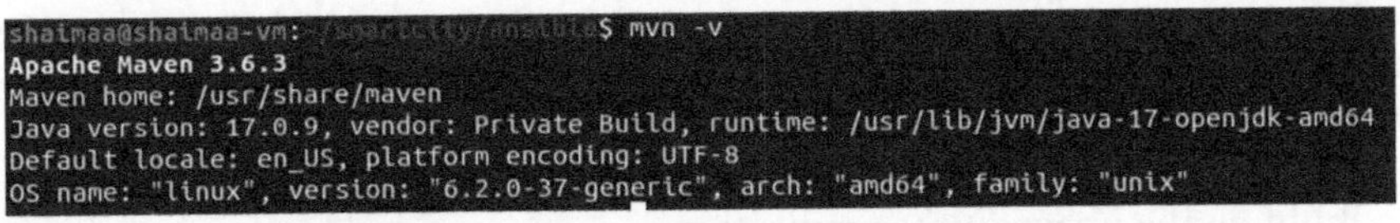

```
shaimaa@shaimaa-vm:~/smartcity/ansible$ mvn -v
Apache Maven 3.6.3
Maven home: /usr/share/maven
Java version: 17.0.9, vendor: Private Build, runtime: /usr/lib/jvm/java-17-openjdk-amd64
Default locale: en_US, platform encoding: UTF-8
OS name: "linux", version: "6.2.0-37-generic", arch: "amd64", family: "unix"
```

Figure 5.12 Installing Maven.

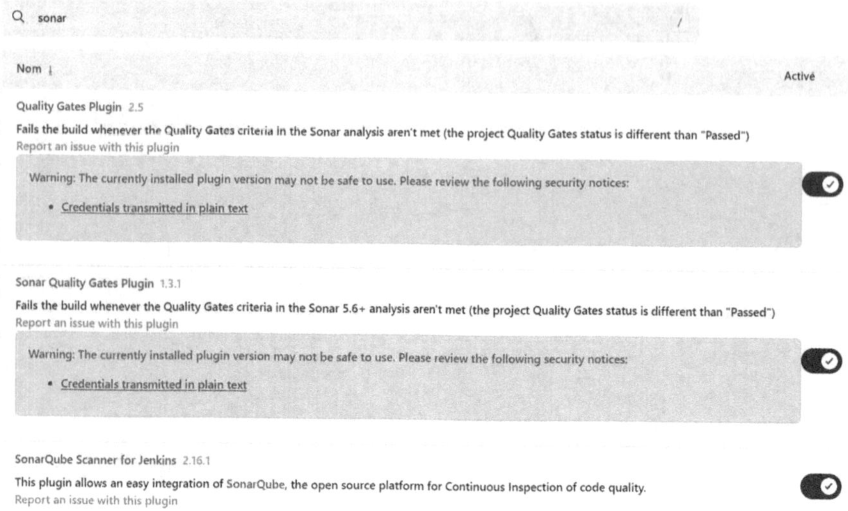

Figure 5.13 Plugin SonarQube.

5.7.7.2 *Integration with Jenkins*

Configure SonarQube Information (Figure 5.14).

5.7.7.3 *Security rule setting*

Open the web interface of your SonarQube server (Figures 5.15 and 5.16).

5.7.7.4 *Checkstyle integration*

Install Checkstyle (Figure 5.17).

5.7.8 Using OWASP Dependency-Check

5.7.8.1 *Plugin installation*

Install the OWASP Dependency-Check plugin for Jenkins to incorporate this analysis into the build process (Figure 5.18).

5.7.9 Docker and Docker Hub

In this chapter, we'll discuss the process of creating a Docker image for your application and uploading it to Docker Hub, a popular Docker container registry (Figure 5.19).

Building the image of the application
Creating a Dockerfile

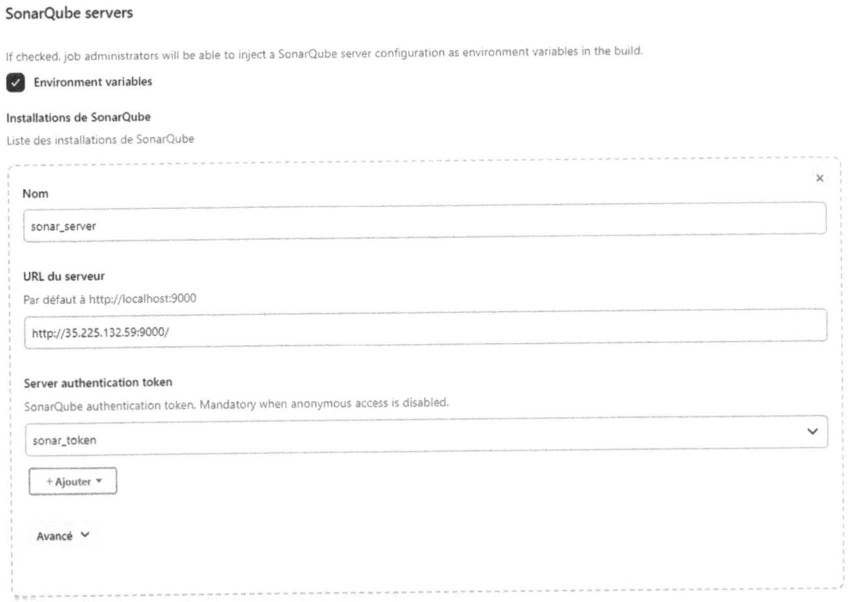

SonarQube servers
If checked, job administrators will be able to inject a SonarQube server configuration as environment variables in the build.
Environment variables
Installations de SonarQube
Liste des installations de SonarQube
Nom
sonar_server
URL du serveur
Par défaut à http://localhost:9000
http://35.225.132.59:9000/
Server authentication token
SonarQube authentication token. Mandatory when anonymous access is disabled.
sonar_token
+ Ajouter
Avancé

Figure 5.14 Configuring the SonarQube server.

Non sécurisé 35.225.132.59:9000/projects
sonarqube
Projects
Issues
Rules
Quality Profiles
Quality Gates
Administration
More
Create Project
My Favorites
All
Filters
Quality Gate
Passed
Failed
Reliability
Security
Perspective
Overall Status
Sort by
Name
1 project(s)
DevSecOps_project
PUBLIC
Passed
Bugs
Vulnerabilities
Hotspots Reviewed
Code Smells
Coverage
Duplications
1 of 1 shown

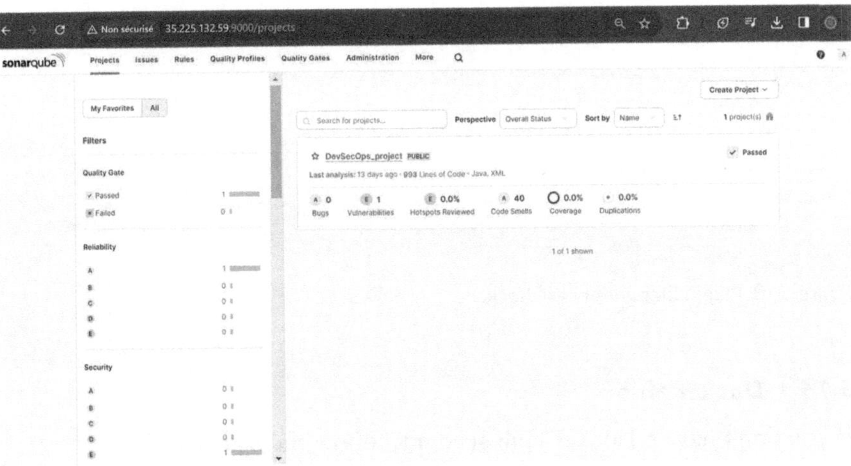

Figure 5.15 SonarQube interface navigate to rule management.

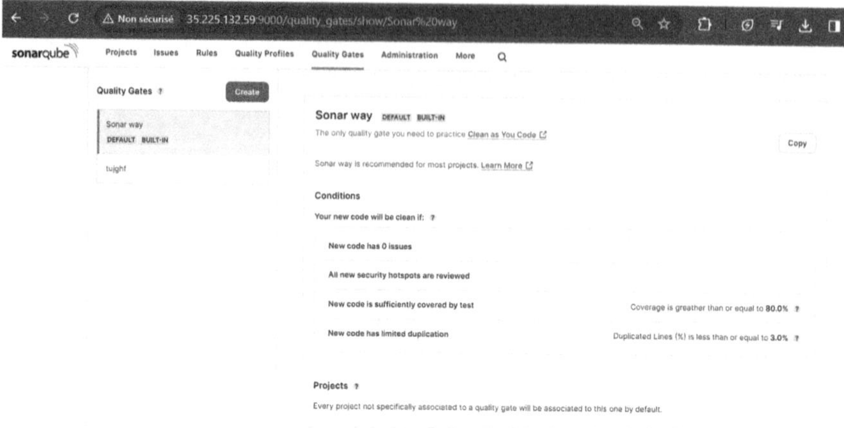

Figure 5.16 Interface SonarQube.

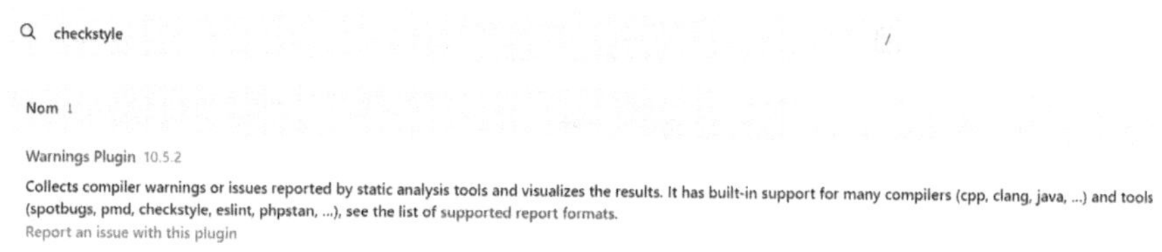

Figure 5.17 Plugin Checkstyle.

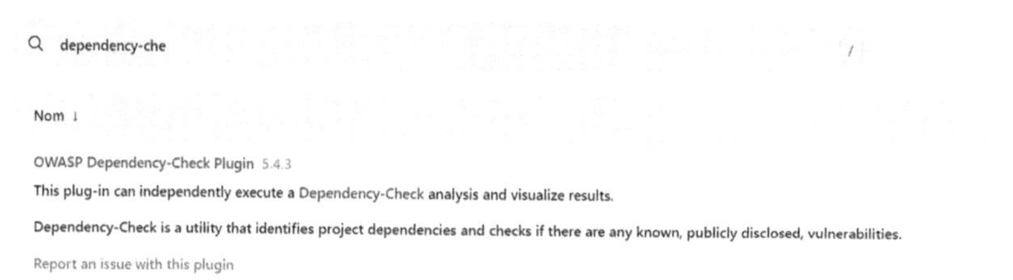

Figure 5.18 Plugin Dependency-Check.

5.7.9.1 *Docker Hub*

If you don't have a Docker Hub account, follow the below steps:

Creating an Account: Visit https://hub.docker.com/.
Click the "Sign Up" button to create a new Docker Hub account. Fill out the Form: Fill out the registration form with the necessary information, including your username, email address, and password.

Dockerfile 457 B

Blame | Edit | Replace | Delete

```
1  # Use the official Maven image to build the application
2  FROM maven:3.8-openjdk-17 AS builder
3
4  WORKDIR /app
5  COPY . .
6  RUN mvn clean
7  RUN mvn install -DskipTests
8
9  # Use the official OpenJDK image for the runtime container
10 FROM openjdk:17-jdk-slim
11
12 WORKDIR /app
13 COPY --from=builder /app/target/smartcity-0.0.1-SNAPSHOT.jar smartcity-app.jar
14
15 # Expose the port your Spring Boot application listens on
16 EXPOSE 8080
17
18 ENTRYPOINT ["java", "-jar", "smartcity-app.jar"]
19
```

Figure 5.19 Dockerfile.

You may need to validate your email address by clicking the confirmation link sent to your inbox.

5.7.9.2 *Jenkins pipeline*

This section will explore setting up a Jenkins pipeline to automate the build and deployment process. We will discuss the Jenkins project, integration with GitLab, the rules, the use of the JenkinsFile, the launch of the pipeline build, and the results obtained (Figure 5.20).

Jenkins project
Integration with GitLab
Installing the plugin

5.7.9.3 *Rules*

Define rules to control pipeline behavior and use rules to trigger specific steps based on branches, labels, or other conditions (Figures 5.21 and 5.22).

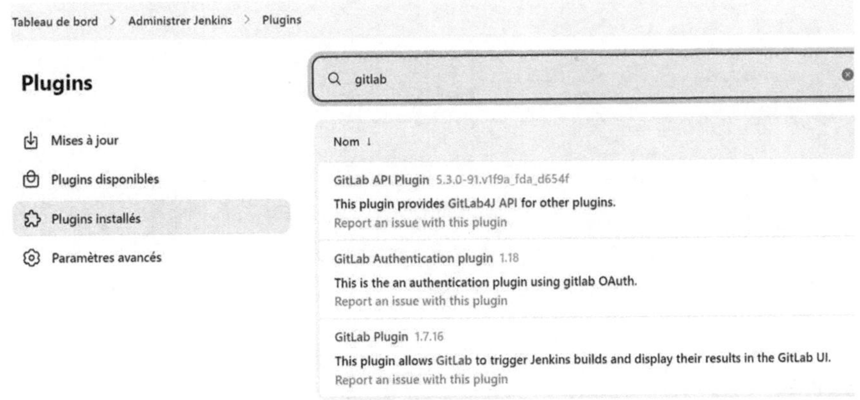

Figure 5.20 Plugin Dependency-Check.

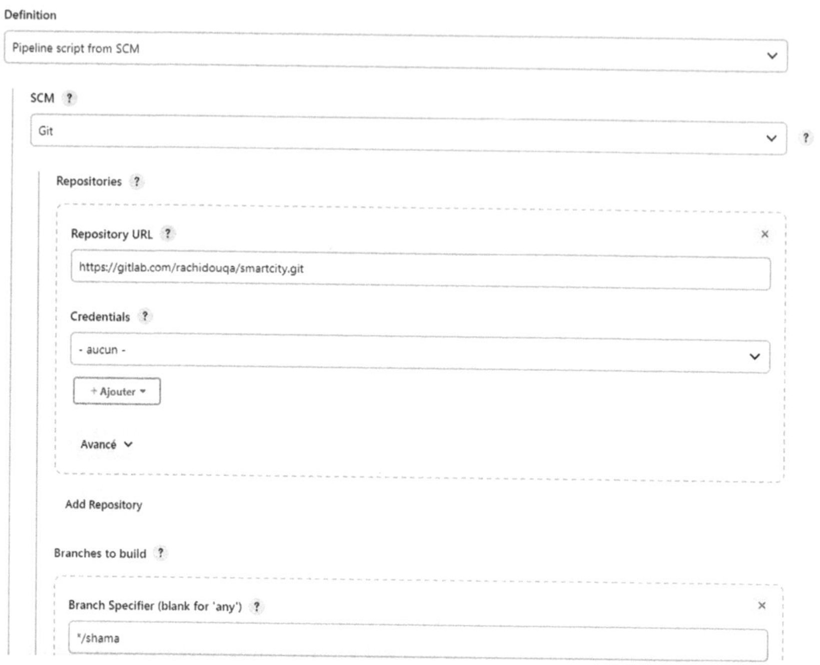

Definition
Pipeline script from SCM
SCM
Git
Repositories
Repository URL
https://gitlab.com/rachidouqa/smartcity.git
Credentials
- aucun -
+ Ajouter
Avancé
Add Repository
Branches to build
Branch Specifier (blank for 'any')
*/shama

Figure 5.21 Configuration.

Using the JenkinsFile
Select the path of the JenkinsFile
Launching the pipeline build

Jenkins offers the ability to manually untrigger the build from its web interface. You can locate and click on the "Build Now" link to start the build process by navigating to the main project page.

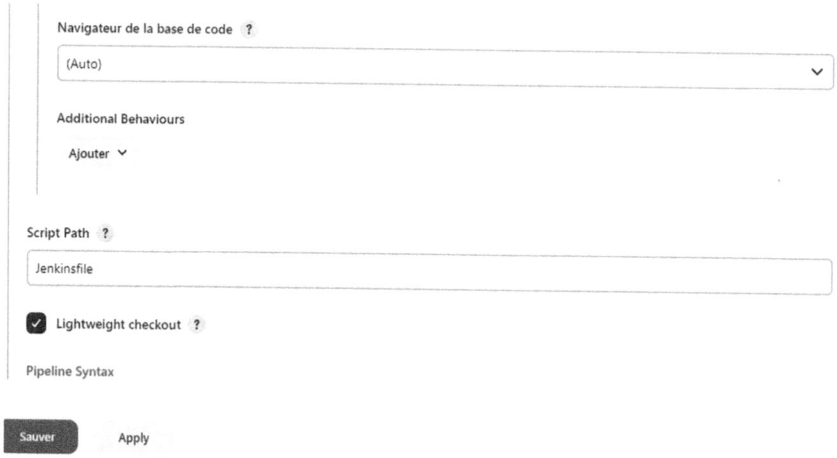

Navigateur de la base de code
(Auto)
Additional Behaviours
Ajouter
Script Path
Jenkinsfile
Lightweight checkout
Pipeline Syntax
Sauver
Apply

Figure 5.22 Path to JenkinsFile.

5.7.9.4 *Result*

Analysis of the results of the various steps, including build, testing, and other actions defined in the JenkinsFile (Figures 5.23–5.27).

5.7.10 Deployment on GKE with Argo CD

This section will detail how to deploy an application on GKE using Argo CD. This includes installing Argo CD, configuring the application, automatically synchronizing, and reviewing the results.

5.7.10.1 *Installing Argo CD*

The installation of Argo CD is the first crucial step in the automated delivery of applications to GKE. This process typically involves deploying Argo CD to the GKE cluster using YAML configuration files (Figures 5.28 and 5.29).

```
[Pipeline] }
[Pipeline] // stage
[Pipeline] withEnv
[Pipeline] {
[Pipeline] stage
[Pipeline] { (Checkout)
[Pipeline] checkout
The recommended git tool is: git
No credentials specified
 > git rev-parse --resolve-git-dir /var/lib/jenkins/workspace/DevSecOps_project/.git # timeout=10
Fetching changes from the remote Git repository
 > git config remote.origin.url https://gitlab.com/rachidouqa/smartcity.git # timeout=10
Fetching upstream changes from https://gitlab.com/rachidouqa/smartcity.git
 > git --version # timeout=10
 > git --version # 'git version 2.34.1'
 > git fetch --tags --force --progress -- https://gitlab.com/rachidouqa/smartcity.git +refs/heads/*:refs/remotes/origin/* # timeout=10
 > git rev-parse refs/remotes/origin/shama^{commit} # timeout=10
Checking out Revision 2d7aa86cd698b0b8081b7fdfda2919e04e1eaec9 (refs/remotes/origin/shama)
 > git config core.sparsecheckout # timeout=10
 > git checkout -f 2d7aa86cd698b0b8081b7fdfda2919e04e1eaec9 # timeout=10
Commit message: "Update Jenkinsfile"
[Pipeline] sh
+ echo Checkout passed
Checkout passed
```

Figure 5.23 Logs of checkout.

```
[□[1;34mINFO□[m]
[□[1;34mINFO□[m] Results:
[□[1;34mINFO□[m]
[□[1;34mINFO□[m] Tests run: 0, Failures: 0, Errors: 0, Skipped: 0
[□[1;34mINFO□[m]
[□[1;34mINFO□[m] □[1m------------------------------------------------------------------------□[m
[□[1;34mINFO□[m] □[1;32mBUILD SUCCESS□[m
[□[1;34mINFO□[m] □[1m------------------------------------------------------------------------□[m
[□[1;34mINFO□[m] Total time:  3.416 s
[□[1;34mINFO□[m] Finished at: 2023-12-23T18:15:56Z
[□[1;34mINFO□[m] □[1m------------------------------------------------------------------------□[m
```

Figure 5.24 Build logs.

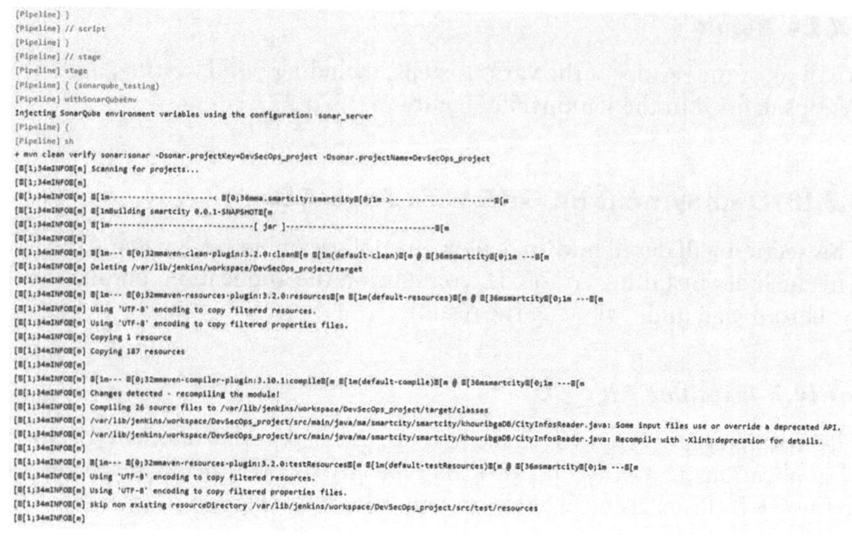

Figure 5.25 Logs of SonarQube.

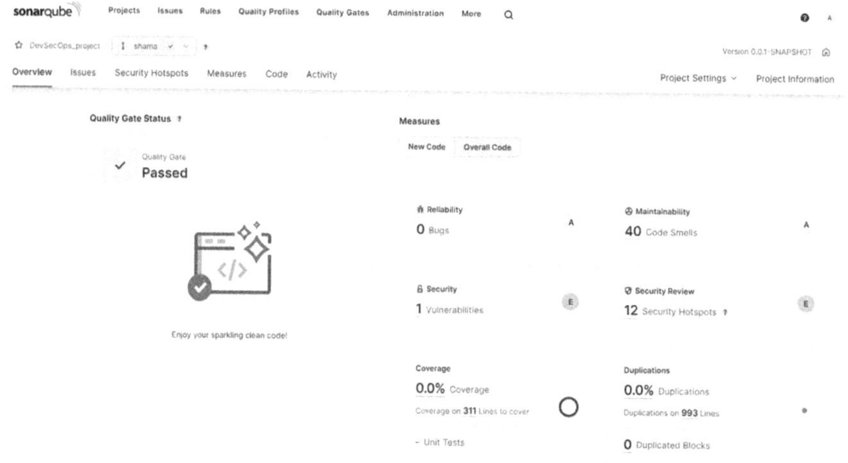

Figure 5.26 Dashboard SonarQube.

5.7.10.2 Setting up the app

Adding an Application to Argo CD

Go to the Argo CD web interface and add a new application by providing details of your application, such as name, Git repository, and path to Kubernetes devices.

Automatic synchronization

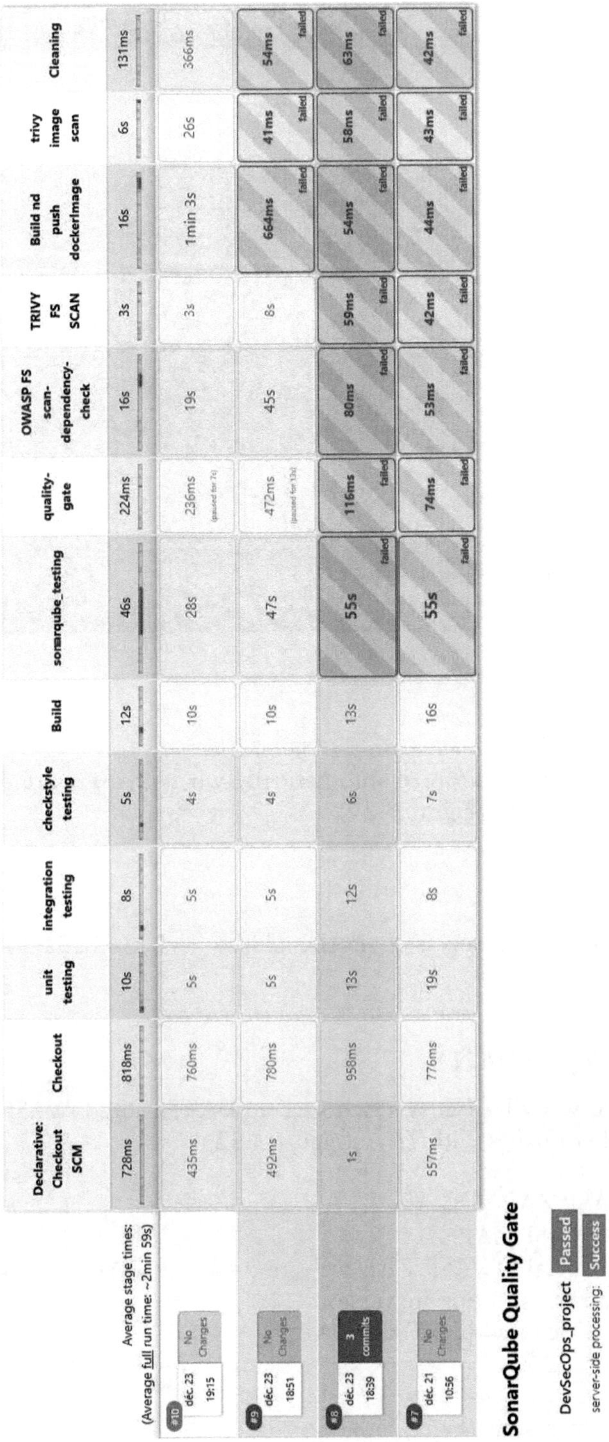

	Declarative: Checkout SCM	Checkout	unit testing	integration testing	checkstyle testing	Build	sonarqube_testing	quality-gate	OWASP FS scan-dependency-check	TRIVY FS SCAN	Build nd push dockerImage	trivy image scan	Cleaning
Average stage times: (Average full run time: ~2min 59s)	728ms	818ms	10s	8s	5s	12s	46s	224ms	16s	3s	16s	6s	131ms
#10 déc. 23 19:15 No Changes	435ms	760ms	5s	5s	4s	10s	28s	236ms (paused for 7s)	19s	3s	1min 3s	26s	366ms
#9 déc. 23 18:51 No Changes	492ms	780ms	5s	5s	4s	10s	47s	472ms (paused for 13s)	45s	8s	664ms failed	41ms failed	54ms failed
#8 déc. 23 18:39 3 commits	1s	958ms	13s	12s	6s	13s	55s failed	116ms failed	80ms failed	59ms failed	54ms failed	58ms failed	63ms failed
#7 déc. 21 10:56 No Changes	557ms	776ms	19s	8s	7s	16s	55s failed	74ms failed	53ms failed	42ms failed	44ms failed	43ms failed	42ms failed

SonarQube Quality Gate

DevSecOps_project Passed

server-side processing: Success

Figure 5.27 Final result.

```
shaimaa@shaimaa-vm:~/smartcity$ kubectl create namespace argocd
namespace/argocd created
```

Figure 5.28 Create namespace.

```
shaimaa@shaimaa-vm:~         $ kubectl apply -n argocd -f https://raw.githubusercontent.com/argoproj/argo-cd/stable/manifests/install.yaml
customresourcedefinition.apiextensions.k8s.io/applications.argoproj.io created
customresourcedefinition.apiextensions.k8s.io/applicationsets.argoproj.io created
customresourcedefinition.apiextensions.k8s.io/appprojects.argoproj.io created
serviceaccount/argocd-application-controller created
serviceaccount/argocd-applicationset-controller created
serviceaccount/argocd-dex-server created
serviceaccount/argocd-notifications-controller created
serviceaccount/argocd-redis created
serviceaccount/argocd-repo-server created
serviceaccount/argocd-server created
role.rbac.authorization.k8s.io/argocd-application-controller created
role.rbac.authorization.k8s.io/argocd-applicationset-controller created
role.rbac.authorization.k8s.io/argocd-dex-server created
role.rbac.authorization.k8s.io/argocd-notifications-controller created
role.rbac.authorization.k8s.io/argocd-server created
clusterrole.rbac.authorization.k8s.io/argocd-application-controller created
clusterrole.rbac.authorization.k8s.io/argocd-server created
rolebinding.rbac.authorization.k8s.io/argocd-application-controller created
rolebinding.rbac.authorization.k8s.io/argocd-applicationset-controller created
rolebinding.rbac.authorization.k8s.io/argocd-dex-server created
rolebinding.rbac.authorization.k8s.io/argocd-notifications-controller created
rolebinding.rbac.authorization.k8s.io/argocd-server created
clusterrolebinding.rbac.authorization.k8s.io/argocd-application-controller created
clusterrolebinding.rbac.authorization.k8s.io/argocd-server created
configmap/argocd-cm created
configmap/argocd-cmd-params-cm created
configmap/argocd-gpg-keys-cm created
configmap/argocd-notifications-cm created
configmap/argocd-rbac-cm created
configmap/argocd-ssh-known-hosts-cm created
configmap/argocd-tls-certs-cm created
secret/argocd-notifications-secret created
secret/argocd-secret created
service/argocd-applicationset-controller created
service/argocd-dex-server created
service/argocd-metrics created
service/argocd-notifications-controller-metrics created
```

Figure 5.29 Application of manifests.

Set Argo CD to synchronize automatically when changes are detected in the Git repository (Figure 5.30).

5.7.10.3 Result

The application can be accessed via the address "34.134.7232:30001." (See Figure 5.31.)

5.7.10.4 Setting up DAST

In this section, we will discuss how to set up DAST using OWASP ZAP and securing Docker images with Trivy (Figure 5.32).

Using OWASP ZAP
Installing OWASP ZAP
Go to the official OWASP ZAP website (*OWASP ZAP Downloads*) and download the appropriate version
Performing a Scan with OWASP ZAP
Launch OWASP ZAP

Analysis of the results (Figure 5.33).

Figure 5.30 The Argo CD application.

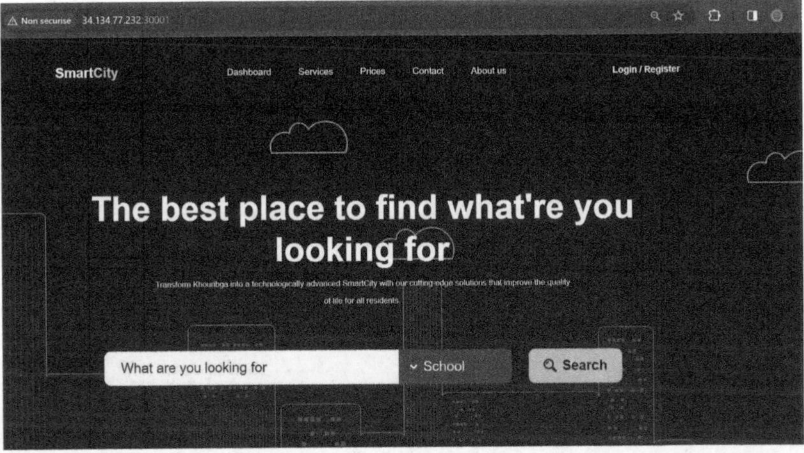

Figure 5.31 The Smartcity app.

5.7.10.5 *Securing images with Trivy*

Installing Trivy
Follow the Trivy installation instructions in the documentation
Docker Image Analysis with Trivy

The image analysis of our avev trivy application (Figure 5.34).

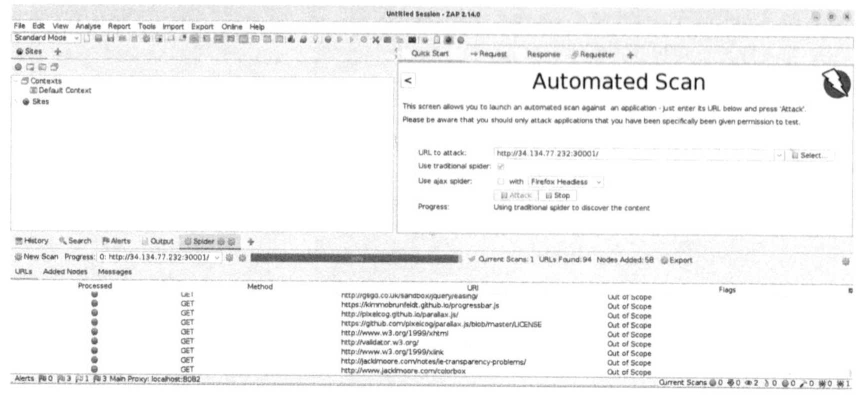

Figure 5.32 OWASP ZAP.

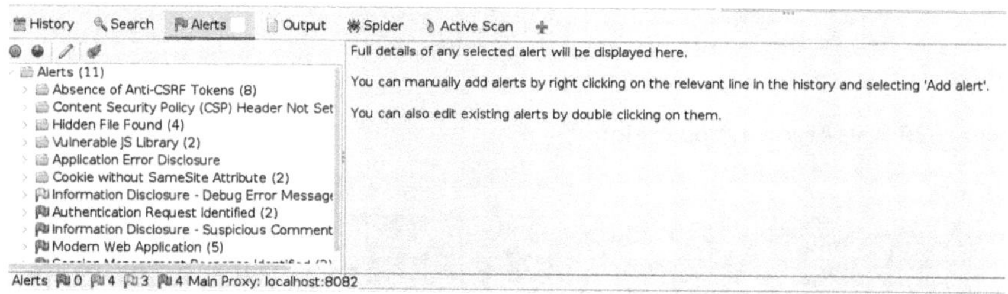

Figure 5.33 Les alerts.

LIBRARY	VULNERABILITY ID	SEVERITY	INSTALLED VERSION	FIXED VERSION	TITLE
apt	CVE-2011-3374	LOW	2.2.4		It was found that apt-key in apt, all versions, do not correctly... -->avd.aquasec.com/nvd/cve-2011-3374
bash	CVE-2022-3715	HIGH	5.1-2		bash: a heap-buffer-overflow in valid_parameter_transform -->avd.aquasec.com/nvd/cve-2022-3715
bsdutils	CVE-2022-0563	MEDIUM	2.36.1-8+deb11u1		util-linux: partial disclosure of arbitrary files in chfn and chsh when compiled... -->avd.aquasec.com/nvd/cve-2022-0563
coreutils	CVE-2016-2781		8.32-4		coreutils: Non-privileged session can escape to the parent session in chroot -->avd.aquasec.com/nvd/cve-2016-2781
	CVE-2017-18018				coreutils: race condition vulnerability in chown and chgrp -->avd.aquasec.com/nvd/cve-2017-18018
dpkg	CVE-2022-1664	CRITICAL	1.20.9	1.20.10	Dagster-cloud 1.1.4 updates 'dagster/dagster-cloud-agent' Docker image's base to 'python:3.8.15-slim' to include security... -->avd.aquasec.com/nvd/cve-2022-1664

Figure 5.34 Trivy scan result.

5.7.11 Cluster monitoring

Now, we will explore how to set up effective monitoring of the Kubernetes cluster on GKE using Prometheus and Grafana. Key steps include installing Prometheus and Grafana, setting up Prometheus as a data source for Grafana, and creating a dashboard to visualize the cluster's metrics.

Installation of the Prometheus and Grafana on GKE

Installing Prometheus (Figures 5.35 and 5.36)
Configuring Prometheus as Grafana's Data Source
Adding Prometheus as a Data Source (Figure 5.37)

5.7.12 Creating the dashboard

Creating a dashboard in Grafana is the final step in visualizing the cluster's metrics in a user-friendly way. This isn't just about getting the most out of the cluster; it's about making that data actionable. With Grafana, you can add panels to display specific metrics such as resource utilization, pod status, and other vital information for cluster monitoring. Customize the dashboard by

```
$ helm install -n monitoring prometheus prometheus-community/prometheus
NAME: prometheus
LAST DEPLOYED: Fri Dec 22 12:28:20 2023
NAMESPACE: monitoring
STATUS: deployed
REVISION: 1
TEST SUITE: None
NOTES:
The Prometheus server can be accessed via port 80 on the following DNS name from within your cluster:
prometheus-server.monitoring.svc.cluster.local


Get the Prometheus server URL by running these commands in the same shell:
  export POD_NAME=$(kubectl get pods --namespace monitoring -l "app.kubernetes.io/name=prometheus,app.kubernetes.io/instance=prometheus" -o jsonpath="{.items[0].metadata.
name}")
  kubectl --namespace monitoring port-forward $POD_NAME 9090


The Prometheus alertmanager can be accessed via port 9093 on the following DNS name from within your cluster:
prometheus-alertmanager.monitoring.svc.cluster.local


Get the Alertmanager URL by running these commands in the same shell:
  export POD_NAME=$(kubectl get pods --namespace monitoring -l "app.kubernetes.io/name=alertmanager,app.kubernetes.io/instance=prometheus" -o jsonpath="{.items[0].metadat
a.name}")
  kubectl --namespace monitoring port-forward $POD_NAME 9093
#################################################################################
######   WARNING: Pod Security Policy has been disabled by default since     #####
######            it deprecated after k8s 1.25+. use                         #####
######            (index .Values "prometheus-node-exporter" "rbac"           #####
######            "pspEnabled") with (index .Values                          #####
######            "prometheus-node-exporter" "rbac" "pspAnnotations")        #####
######            in case you still need it.                                 #####
#################################################################################


The Prometheus PushGateway can be accessed via port 9091 on the following DNS name from within your cluster:
prometheus-prometheus-pushgateway.monitoring.svc.cluster.local
```

Figure 5.35 Installing Prometheus installing Grafana.

```
$ helm install grafana grafana/grafana -n monitoring
NAME: grafana
LAST DEPLOYED: Fri Dec 22 12:31:23 2023
NAMESPACE: monitoring
STATUS: deployed
REVISION: 1
NOTES:
1. Get your 'admin' user password by running:

   kubectl get secret --namespace monitoring grafana -o jsonpath="{.data.admin-password}" | base64 --decode ; echo


2. The Grafana server can be accessed via port 80 on the following DNS name from within your cluster:

   grafana.monitoring.svc.cluster.local

   Get the Grafana URL to visit by running these commands in the same shell:
     export POD_NAME=$(kubectl get pods --namespace monitoring -l "app.kubernetes.io/name=grafana,app.kubernetes.io/instance=grafana" -o jsonpath="{.items[0].metadata.nam
e}")
     kubectl --namespace monitoring port-forward $POD_NAME 3000

3. Login with the password from step 1 and the username: admin
#################################################################################
######   WARNING: Persistence is disabled!!! You will lose your data when   #####
######            the Grafana pod is terminated.                            #####
#################################################################################
```

Figure 5.36 Installation of Prometheus.

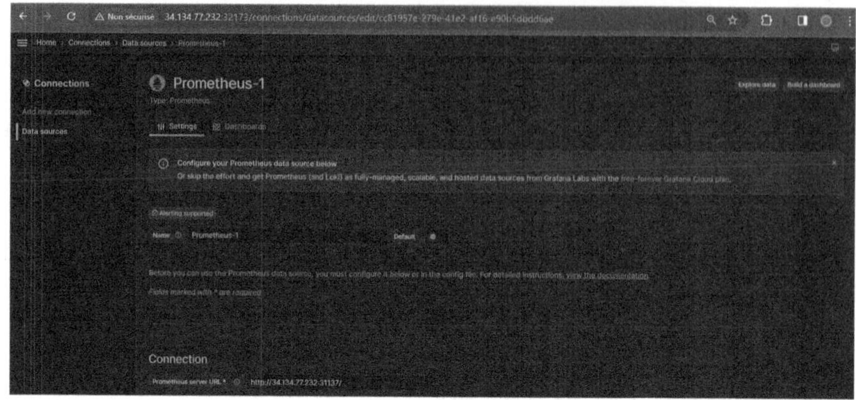

Figure 5.37 Installing Prometheus.

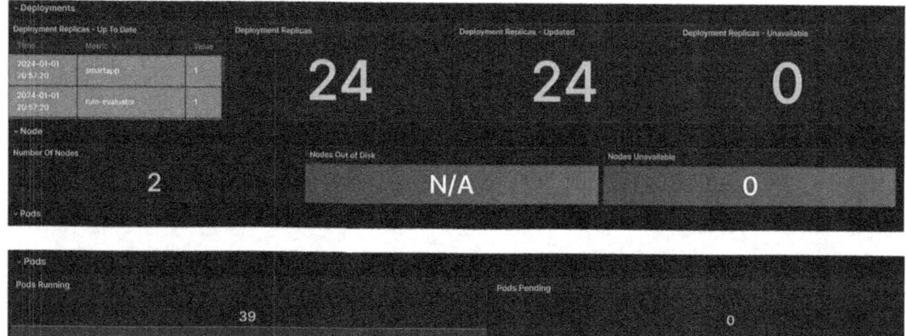

Figure 5.38 Dashboard Grafana.

adding annotations, filters, and other graphical elements to gain an in-depth understanding of the cluster's status (Figure 5.38).

5.8 CONCLUSION

The integration of DevSecOps within Agile environments represents a significant advancement in the field of cybersecurity, addressing the urgent need for continuous, integrated security practices in web application development. This chapter has detailed the methodologies, lifecycle, and practical applications of DevSecOps, highlighting its necessity and efficiency in today's digital landscape. By implementing DevSecOps, organizations can not only expedite development processes but also enhance their security posture, effectively reducing vulnerabilities and the potential for cyber incidents. The use case presented reaffirms that DevSecOps is not merely a theoretical model but a practical, indispensable strategy that aligns with and supports

Agile methodologies. As cyber threats continue to evolve, so too must our approaches to developing secure software, making DevSecOps an essential practice for any organization committed to reliable, secure software delivery.

REFERENCES

Abdelkebir, S., Maleh, Y., & Belaissaoui, M. (2019). Towards an Agile and Secure IT Service Management. In *Global Information Diffusion and Management in Contemporary Society* (pp. 125–152). IGI Global.

Díaz, J., Pérez, J. E., Lopez-Peña, M. A., Mena, G. A., & Yagüe, A. (2019). Self-service cybersecurity monitoring as enabler for DevSecOps. *IEEE Access*, *7*, 100283–100295. https://doi.org/10.1109/ACCESS.2019.2930000

Ebert, C., Gallardo, G., Hernantes, J., & Serrano, N. (2016). DevOps. *IEEE Software*, *33*(3), 94–100. https://doi.org/10.1109/MS.2016.68

Gikas, C. (2010). A general comparison of FISMA, HIPAA, ISO 27000 and PCI-DSS standards. *Information Security Journal: A Global Perspective*, *19*(3), 132–141. https://doi.org/10.1080/19393551003657019

HIPAA. (1999). *HIPAA Health Insurance Portability and Accountability Act of 1999*. Http://Www.Hhs.Gov/Ocr/Privacy/Hipaa/Administrative/Privacyrule.

Hsu, T. H. C. (2018). *Hands-On Security in DevOps: Ensure Continuous Security, Deployment, and Delivery With DevSecOps*. Packt Publishing Ltd.

Humphreys, E. (2008). Information security management standards: Compliance, governance and risk management. *Information Security Technical Report*, *13*(4), 247–255. https://doi.org/http://dx.doi.org/10.1016/j.istr.2008.10.010

Jakobsson, A., & Häggström, I. (2022). *Study of the Techniques Used by OWASP ZAP for Analysis of Vulnerabilities in Web Applications*.

Li, J. (2020). Vulnerabilities mapping based on OWASP-SANS: A survey for static application security testing (SAST). *ArXiv Preprint ArXiv:2004.03216*.

Patel, V., Mohandas, R., & Pais, A. R. (2010). Attacks on Web Services and Mitigation Schemes. *Security and Cryptography (SECRYPT), Proceedings of the 2010 International Conference On* (pp. 1–6).

Prates, L., Faustino, J., Silva, M., & Pereira, R. (2019). *DevSecOps Metrics BT - Information Systems: Research, Development, Applications, Education* (S. Wrycza & J. Maślankowski, Eds.; pp. 77–90). Springer International Publishing.

Read, W., Report, T., & Takeaways, K. (2016). *Agile And DevOps Adoption Drives Digital Business Success*.

Sadqi, Y., & Maleh, Y. (2022). A systematic review and taxonomy of web applications threats. *Information Security Journal: A Global Perspective*, *31*(1), 1–27. https://doi.org/10.1080/19393555.2020.1853855

Sahid, A., Maleh, Y., & Belaissaoui, M. (2020a). Strategic Information System Agility: From Theory to Practices. In *Strategic Information System Agility: From Theory to Practices*. Emerald Publishing Limited. https://doi.org/10.1108/978-1-80043-810-120211001

Sahid, A., Maleh, Y., & Belaissaoui, M. (2020b). Understanding Agility Concept. In *Strategic Information System Agility: From Theory to Practices* (pp. 9–27). Emerald Publishing Limited. https://doi.org/10.1108/978-1-80043-810-120211003

Sahid, A., Maleh, Y., & Belaissoui, M. (n.d.). *A Practical Agile Framework for IT Assets Management: A Mixed Methods Study*.

Yassine, M., Abdelkebir, S., & Abdellah, E. (2017). A Capability Maturity Framework for IT Security Governance in Organizations. *13th International Symposium on Information Assurance and Security (IAS 17)*.

Chapter 6

Penetration testing in agile cybersecurity environments

Abdullah S. Alshra'a, Loui Al Sardy, and Reinhard German

6.1 INTRODUCTION

6.1.1 Overview of agile security

Agile security is a rapidly growing cybersecurity approach due to its ability to quickly adapt and respond to changing threats and environments [1, 2]. As the complexity and frequency of cyber-attacks increase, security professionals have begun to adopt agile principles to increase the efficiency and effectiveness of their security efforts [3]. By using agile concepts such as continuous development (CD), frequent testing, and collaboration, security teams can detect and mitigate vulnerabilities in their systems [4].

Agile security improves the resilience and responsiveness of organizations against cyber threats by including agile principles such as continuous monitoring, rapid feedback, and automated testing [5]. Agile security correspondingly offers numerous benefits such as increased agility and faster response times. Thus, security teams are better equipped to identify and address vulnerabilities in real environments and can effectively minimize the risk of data breaches and other security incidents [6].

The research methodology used in this chapter adopts a qualitative approach, focusing on thoroughly exploring agile security and penetration testing concepts. The chapter conducts an in-depth review of the existing literature to examine the theoretical foundations of agile security, including its principles, practices, and significance in today's digital environments. By Strengthen knowledge from academic journals, industry reports, and expert publications, the methodology aims to establish a strong understanding of integrating security practices within agile development frameworks. Additionally, the chapter aims to clarify the crucial role of Penetration Testing in agile security by analyzing its types, components, and methodologies, providing insights into how organizations can effectively protect their digital assets amid evolving cybersecurity threats.

In addition to the theoretical exploration, the research methodology includes practical insights from case studies and real-world examples. By analyzing the challenges faced during the implementation of penetration

 DOI: 10.1201/9781003478676-6

testing in agile environments, the chapter aims to offer practical solutions and best practices for overcoming these obstacles. Through a qualitative investigation into the dynamics of red teams, blue teams, and the emerging purple teams, the methodology sheds light on the complexities of team collaboration and communication in agile security contexts. By combining theoretical frameworks with practical experiences, the research methodology aims to provide a comprehensive understanding of agile security and penetration testing, paving the way for future research directions and industry advancements.

This chapter details the importance of penetration testing in agile security. It starts by introducing the principles and methodologies of agile security, along with an overview of the pivotal role played by penetration testing in ensuring security within Agile frameworks. The chapter then delves into the challenges posed by digital security threats, setting the stage for a comprehensive discussion on the various facets of penetration testing in Agile environments. This includes an examination of the different types and components of penetration testing and how they align with agile security practices.

Moving forward, the chapter explores penetration testing methodologies, including the roles and responsibilities of team members, the unique challenges encountered in Agile settings, and the different classifications of penetration testing. It also looks at the crucial task of assembling the right team for agile security, considering dynamics such as the red team vs. blue team approaches, the emergence of the purple team concept, and the distinctions between internal and external team dynamics.

Furthermore, the chapter delves into the nuances of penetration testing, including understanding its underlying principles and deciphering hacker intentions. It addresses the specific challenges encountered during penetration testing in agile security and proposes effective solutions to mitigate these challenges, emphasizing the importance of communication and collaboration.

Finally, the chapter navigates through the best practices in Agile penetration testing and outlines potential future research directions in this evolving field. It concludes by encapsulating the key findings and insights presented throughout the chapter.

6.1.2 Importance of penetration testing in agile security

The fundamental principle of agile security appears through continuous improvement, where the security measures are constantly evaluated and repeated to adapt to changing threats and vulnerabilities [7]. Penetration testing [8] is an essential element in providing security to the organization's digital assets and supporting agile security by identifying vulnerabilities in an organization's systems and applications, allowing security teams to handle potential weaknesses before being exploited by malicious actors. Therefore, the organization mostly conducts regular penetration tests to avoid cyber criminals and minimize the risk of being victim to costly data breaches or security incidents [9]. Highlighted below are essential points emphasizing the value of penetration testing in agile security:

- Early Detection of Vulnerabilities: Once the organization relies on agile security, the software is developed in short and iterative cycles. Hence, penetration testing as a proactive process ensures that security is not only a reaction but an inherent part of the agile security life cycle. Thus, penetration testing mostly guarantees that security threats are identified before becoming severe problems [10].
- Risk Mitigation in Agile Development: Agile projects are distinguished by continuous adaption to altering requirements. Penetration testing facilitates the adaptation of security measures concurrently and mitigates risks related to evolving project scopes and functionalities. Additionally, teams can make proper decisions to address security concerns by identifying vulnerabilities instantly [11].
- Integration of Security into DevOps Pipeline: DevOps [12] combines Software Development (Dev) and IT Operations (Ops). It improves teamwork and automates the processes of software delivery to faster and more reliable development and deployment cycles. Once the penetration testing tools are integrated into the DevOps pipeline in agile security, the automated security testing boosts ensuring that the security assessments are part of the Continuous Integration and CD (CI/CD) processes [13]. This integration enhances the overall security situation of the software [14].
- Cost-Effective Security Assurance: Identifying and resolving security vulnerabilities during the development phase is more cost-effective than handling their hazards post-release. Penetration testing decreases the overall cost of security by identifying and treating these issues early in the development life cycle, in addition to preventing costly security incidents later on [15].
- Regulatory Compliance and Stakeholder Trust: Many industries have rigid organizational requirements for security. Accordingly, penetration testing ensures that the development teams fulfill these compliance standards by regularly estimating and validating the security controls, and hence makes trust among stakeholders, customers, and regulatory elements [16].
- Endless Security Enhancements: Agile security boosts a culture of continuousadvancement by frequent penetration testing that provides opportunities for teams to understand security findings, implement corrective measurements, and enhance their security conditions and overall stability [17].

Ultimately, penetration testing is an essential element of a comprehensive cybersecurity technique, helping organizations stay ahead of potential threats and strengthen their protection. Proactive identifying and remediating vulnerabilities support companies in lowering the risk of security breaches, protecting sensitive data, and establishing a commitment to maintaining a secure digital environment [18].

6.1.3 Growth of digital security challenges

As long as technology advances outstandingly, the cybersecurity landscape continues to shift and presents new challenges that require creative solutions [19, 20]. Understanding the evolution of these challenges is important for organizations to protect their assets in an increasingly interconnected and complex environment. Figure 6.1 illustrates a range of cyber security attack vectors, covering a spectrum of threats observed in modern digital environments. At the beginning of digitization appearance, security concerns mainly focus on viruses, worms, and malware. So, cybersecurity measures concentrated on defenses against these traditional threats, often using antivirus software and firewalls. But, with coming sophisticated cyber-attacks because of the internet spread and the growth of interconnected systems, cyber-criminals have improved their tactics [21]. Advanced Persistent Threats (APTs) and sophisticated hacking techniques emerged, targeting specific organizations for spying, data theft, or disruption [22].

Correspondingly, the growth of Internet of Things (IoT) devices has immensely raised the attack surface because of its widespread from smart homes to industrial control systems (ICS) [20]. Each connected device represents potential vulnerabilities and different challenges in securing decentralized networks. Furthermore, the widespread of mobile devices present more cybersecurity challenges for protecting sensitive information related to mobile malware, insecure apps, and blending of personal and professional use on mobile platforms [23]. Also, cyber-attackers utilize social engineering methods to exploit human vulnerabilities [24, 25]. Phishing [26], pretexting [27], and other manipulative techniques are exploited to deceive individuals

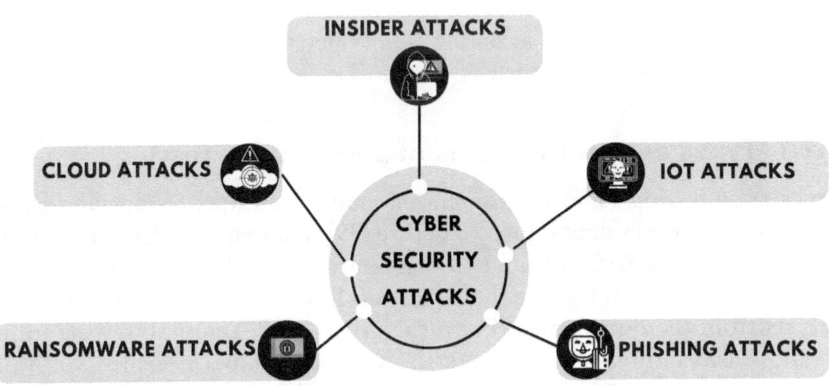

Figure 6.1 Cyber security attacks.

into disclosing sensitive information [28]. Besides, insider threats (whether malicious or unintentional) and ransomware attacks pose a persistent risk. Especially, ransomware attacks [29] become more prevalent and sophisticated, where cyber-criminals not only encrypt data but also employ extortion tactics, threatening to release sensitive information unless a ransom is paid [30].

On the other hand, when many firms move to cloud environments for more scalability and flexibility, securing data stored and processed in cloud computing is a critical issue with shared responsibility models, which require robust access controls and present endless challenges in cloud security [31]. Similarly, Artificial Intelligence (AI) and Machine Learning (ML) present new opportunities for improving security measurements but introduce challenges [32–34]. Cyber-attackers exploit AI for more targeted attacks, and the potential for adversarial attacks on AI systems raises concerns [35].

Generally, the evolution of digital security challenges reflects the dynamic nature of the technological landscape. As organizations adopt emerging technologies, they must continuously adapt their security strategies to address new threats and vulnerabilities. A proactive and agile approach to cybersecurity is essential to stay ahead in the ongoing battle against cyber threats in the digital era. Organizations must navigate complex compliance requirements, and ensure secure personal and sensitive data. Therefore, different guidances for regulatory compliance and data privacy appeared to represent the importance of data protection (e.g., GDPR [36], CCPA [37]).

6.2 PENETRATION TESTING IN AGILE SECURITY: TYPES, COMPONENTS, AND STEPS

With the evolution of cybersecurity, penetration testing appears as a pivotal practice by offering organizations further ability to identify and mitigate vulnerabilities [38]. This section highlights the various types of penetration testing, explores its key components, and summarizes the essential points for its effective execution. Special attention is provided to the integration of penetration testing within agile security frameworks.

6.2.1 Types and components of penetration testing

Penetration testing contains different methods, each suited to specific aspects of an organization's digital infrastructure [39]. For instance, **External Testing** examines attacks arising from external sources outside of the organization's infrastructure to evaluate the robustness of the organization against potential threats from the outside world. On the other hand, **Internal Testing** impersonates insider threats to estimate the security countermeasures within an organization in addition to identifying vulnerabilities that might be exploited by individuals with internal access [40].

In the **Web Application Testing**, the tester concentrates on web-based systems and their databases. Thus, it detects vulnerabilities specific to web applications, which are critical components in today's digital landscape [41]. With the increase of mobile devices, the **Mobile Application Testing** analyzes mobile applications, ensuring their resilience against potential security weaknesses [42]. Also, the **Wireless Network Testing** evaluates the organization's wireless network and assesses the vulnerability of wireless connections to unauthorized access, addressing an essential dimension of modern cybersecurity [43].

All aforementioned types have the same structured approach and a clear definition of their components to show the desired effectiveness described in Figure 6.2 and demonstrated as follows [44, 45]:

1. **Scope Definition:** A precise definition of the testing scope is essential, especially in agile security environments. This includes identifying the systems, networks, or applications to be tested and determining the testing limits.
2. **Rules of Engagement:** The responsible administration establishes rules of engagement and outlines proper actions during testing, ensuring alignment with continuous agile development cycles.
3. **Reconnaissance:** The initial stage includes collecting information regarding the target, understanding its architecture, and identifying

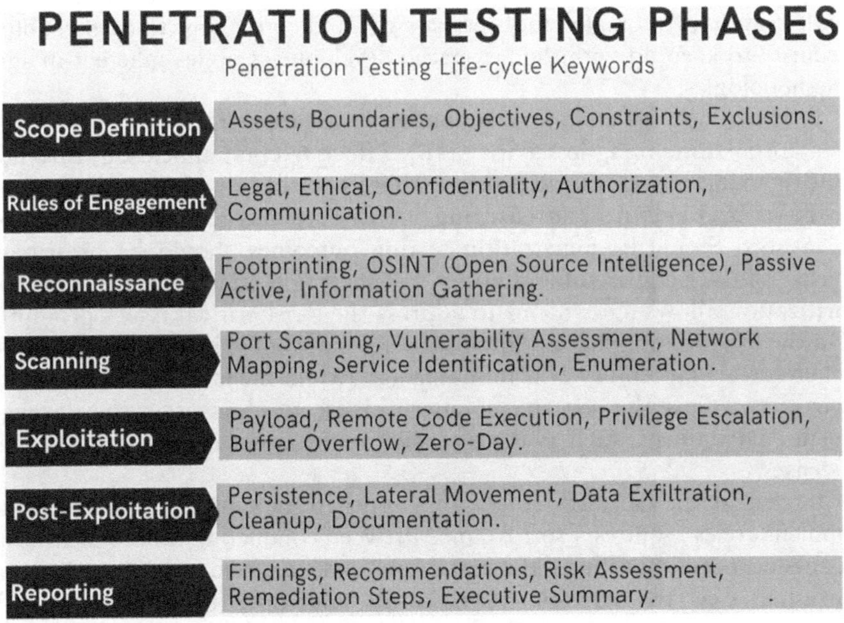

PENETRATION TESTING PHASES

Penetration Testing Life-cycle Keywords

Phase	Keywords
Scope Definition	Assets, Boundaries, Objectives, Constraints, Exclusions.
Rules of Engagement	Legal, Ethical, Confidentiality, Authorization, Communication.
Reconnaissance	Footprinting, OSINT (Open Source Intelligence), Passive Active, Information Gathering.
Scanning	Port Scanning, Vulnerability Assessment, Network Mapping, Service Identification, Enumeration.
Exploitation	Payload, Remote Code Execution, Privilege Escalation, Buffer Overflow, Zero-Day.
Post-Exploitation	Persistence, Lateral Movement, Data Exfiltration, Cleanup, Documentation.
Reporting	Findings, Recommendations, Risk Assessment, Remediation Steps, Executive Summary.

Figure 6.2 Penetration testing phases.

potential entry points. Moreover, this stage should be continuous to adapt to evolving systems which ensures agile security.

4. **Scanning:** Automated tools are employed to scan for vulnerabilities. Integration of automated scanning tools in agile environments ensures real-time identification of potential weaknesses.
5. **Exploitation:** Skilled penetration testers attempt to exploit identified vulnerabilities, simulating real-world cyber-attacks. In agile security, this phase needs to be agile itself, aligning with the rapid development cycles.
6. **Post-Exploitation:** This stage provides rapid feedback for immediate remediation. After successful exploitation, the priority shifts to assessing the extent of the compromise and identifying additional risks.
7. **Reporting:** Thorough reports are generated, detailing identified vulnerabilities, their potential impact, and recommended remediation measures. In agile security, actionable and prioritized insights are crucial for swift mitigation.

6.2.2 Penetration testing phases and their relation to agile security

Penetration testing applies several key considerations and practices that align in the context of agile security [9]. Therefore, in agile environments, penetration testing must be smoothly integrated with the agile security workflow and adopt continuous testing practices, hence ensuring that the rapid pace of development does not violate security measures. To emphasize speed's role in agile security, Automated tools boost swift and continuous testing, enabling security to keep up with the fast-paced development cycles inherent in agile methodologies.

Furthermore, the collaboration between development and security teams is a cornerstone of agile security [46]. Effective communication, throughout the penetration testing process, enabling a deep understanding of the evolving architecture and ensuring the smooth incorporation of security measures. Similarly, penetration testing outcomes should be prioritized based on the rigor of vulnerabilities and their potential influence. This prioritization allows agile teams to address the most critical issues promptly. On the other hand, penetration testing methods must be flexible and agile. Therefore, adaptability is a fundamental requirement in agile security. It works to adjust the security countermeasure to changes in the development environment, such as the introduction of new features, updates, and releases.

In general, the successful integration of penetration testing into agile development cycles requires a soft balance between conducting thorough security assessments and fulfilling the demands of rapid, continuous delivery. As organizations steer the dynamic cybersecurity landscape, the adoption of agile security in penetration testing becomes imperative for building robust and resilient security conditions [47].

6.2.3 Ethical implications of penetration testing in agile environments

Penetration testing is increasingly integrated into agile development environments to identify and mitigate vulnerabilities in software systems. Therefore, when it comes to performing penetration testing in agile environments, the professional needs to have a deep understanding of the ethical implications. This includes challenges related to data privacy, the need for strong data protection measures, and the importance of obtaining informed consent from stakeholders. By addressing these ethical considerations, agile teams control the complex landscape of penetration testing while maintaining ethical standards and safeguarding stakeholders' interests. this integration presents significant ethical concerns with privacy, data protection, and consent, as summarized in the following [48]:

- **Privacy**: Penetration testing often involves collecting and analyzing sensitive information about systems, applications, and networks. In agile environments, where development cycles are rapid and iterative, there's a risk of inadvertently accessing or storing personal data without proper consent or authorization. Thus, defining the scope of penetration tests is crucial to ensure that only authorized systems and assets are targeted. However, in agile environments characterized by dynamic changes and evolving requirements, maintaining an accurate scope is a big challenge, potentially leading to unintended data collection or intrusion into unauthorized systems. The practices of ethical penetration testing emphasize the principle of data minimization, whereby only the minimum amount of data necessary for testing purposes is collected and retained. Agile teams must ensure collecting and handling data responsibly in adhering to privacy regulations and industry best practices [49].
- **Data Protection**: Penetration testing actively searches for vulnerabilities in computer systems, which may accidentally expose sensitive data if the proper security measures are not taken. Therefore, Agile teams must prioritize data security throughout the testing process, implementing encryption, access controls, and other security measures to protect sensitive information from unauthorized access or disclosure. Accordingly, agile development methodologies emphasize iterative development and frequent releases, which may introduce security vulnerabilities if not implemented securely. Therefore, it is important that penetration testing in agile environments not only identifies existing vulnerabilities but also advocates for secure development practices such as threat modeling, code reviews, and secure coding guidelines to prevent future vulnerabilities from arising. Despite proactive security measures, penetration testing may uncover vulnerabilities or incidents that require immediate attention. Therefore, Agile teams must have robust incident response procedures in place to address security incidents promptly, mitigate potential impacts, and ensure compliance with data protection regulations [50].

- **Consent:** Ethical penetration testing requires acquiring informed consent from stakeholders before running tests that may impact their systems or data. However, the frequent obtaining of timely consent from stakeholders in agile environments would be a major challenge due to the fast-paced nature of development cycles. Hence, agile teams must establish clear communication channels with stakeholders and ensure obtaining consent before initiating penetration tests. Also, agile teams should maintain transparency throughout the penetration testing process, informing stakeholders about the purpose, scope, and potential risks associated with the tests. Besides, teams must maintain accountability for their actions, ensuring that penetration testing activities are conducted ethically and in compliance with relevant regulations and ethical guidelines. In agile environments, user acceptance testing (UAT) is often achieved to validate the functionality and usability of software from an end-user perspective. Accordingly, penetration testing should not interfere with UAT activities or compromise user data integrity without stakeholders' explicit consent [51].

6.3 PENETRATION TESTING TEAMS

6.3.1 Team members, roles and responsibilities

In agile security, the success of penetration testing depends not only on well-defined functions but also on the design and collaboration within the testing teams [46]. Therefore, agile security requires a cross-functional and adaptive technique to guarantee that testing aligns seamlessly with the iterative nature of agile development. Accordingly, the penetration testing team comprises four members: Security Experts, Agile Specialists, Developers, and Collaboration Facilitators.

First, the **Security Experts** are the core of a penetration testing team, who have in-depth knowledge of cybersecurity principles, attack vectors, and defensive measures. These professionals fetch expertise in identifying and exploiting vulnerabilities. While the **Agile Specialists** have a strong understanding of agile methodologies to enable penetration testing to integrate effectively into agile workflows. Also, these specialists guarantee that security testing aligns with the principles of continuous integration, continuous delivery, and iterative development. The **Developers** existing in the penetration testing team are useful for a holistic strategy. Their insights into the sophistication of the code-base and application architecture contribute to more effective testing. Likewise, the **Collaboration Facilitators**, who are individuals qualified and skilled in promoting collaboration, ensure effective communication between security experts, developers, and other stakeholders, facilitating a smooth integration of security testing into the agile development life cycle [52, 53].

The four team members collaborate to achieve a set of tasks, where each has various roles depending on their responsibilities, which are summarized and detailed as follows [54–56]:

- **Scrum Master:** Ensure that penetration testing aligns with cycles in an agile context. A Scrum master is a facilitator for an agile development team and is responsible for managing the exchange of information between team members. Scrum is a project management framework that enables a team to communicate and self-organize to make modifications fast according to agile principles. Scrum masters facilitate communication, remove obstacles, and ensure the security testing process does not disrupt the agile workflow.
- **Security Champions:** is a developer with a security interest in helping to amplify the security message at the team level. Security champions just act as the security conscience of the team, keeping their eyes and ears open for potential issues among development teams. Also, having security champions who specialize in secure coding practices is invaluable. These individuals act as links between the development and security teams, promoting security awareness and best practices.
- **Tester-Developers Collaboration:** Close collaboration between penetration testers and developers is essential. Developers activly participate in identifying and remediating vulnerabilities, fostering a shared responsibility for security.
- **Continuous Learning Advocates:** Agile security teams should promote a culture of continuous learning. This involves staying updated on the latest security threats, vulnerabilities, and testing techniques to adapt to the evolving threat landscape.

Based on the above, it is noticeable that the critical role of the penetration testing teams in confirming that security measures align with the principles of agility. However, some challenges need proactive resolutions that contribute to the effective integration of penetration testing into agile development cycles [9]. For instance, the time constraints challenge is inherent in agile environments. Hence, penetration testing teams are required to employ efficient testing methodologies, leverage automation, and prioritize high-impact vulnerabilities to address this challenge.

Similarly, the resolution of communication gaps is essential to agile security to keep effective communication. Regular stand-up meetings, collaboration tools, and shared documentation support bridging communication gaps between team members and stakeholders. The penetration testing team must take into consideration balancing speed and security as a persistent challenge. The team must strike a balance between the speed of agile development in addition to the accuracy and thoroughness of security testing. Regular retrospectives and process improvements help fine-tune the integration of security testing into agile cycles.

A well-rounded and collaborative team composition, clear roles and responsibilities, and proactive solutions to challenges contribute to the effective integration of penetration testing into agile development cycles. By fostering collaboration and continuous learning, penetration testing teams become key enablers of resilient and secure software development practices [46].

6.3.2 Classification of penetration testing

Penetration testing is often conducted by specialized teams, each with distinct roles, expertise, and ethical considerations. These teams play a crucial role in ensuring the security of agile environments. Penetration testing teams are commonly classified based on the level of knowledge they possess about the target system and the extent of information provided to them:

- **White Box Testing [57, 58]:** Also known as clear box or glass box testing, it involves penetration testers having a full understanding of the target system and having access to the source code and environment. The target of white box testing is to perform an in-depth security audit of a business's systems and to provide the testers with as much detail as possible. As a result, the tests are more thorough because the tester has access to areas and then can simulate from an insider's perspective which allows the team to explore potential vulnerabilities with in-depth understanding. Nevertheless, White box tests do have their drawbacks. For example, if the pen tester has a high level of access, that would lead to taking a longer time to decide what areas to focus on. In addition, the testing often demands sophisticated and expensive tools such as code analyzers and debuggers.
- **Black Box Testing [58, 59]:** In contrast, black box testing operates with minimal knowledge of the target system and the IT infrastructure of a business. The main advantage of this method of testing is to simulate a real-world cyber-attack, whereby the pen tester supposes the role of an uninformed attacker. Testers test the system as external attackers would, with no prior information about the internal infrastructure. This type of testing provides a more realistic assessment of how a real-world attacker might exploit the system. During a black box penetration test, the pen tester is given little to no information. Black Box Testing is also classified as the "trial and error" technique and needs a high degree of technical skill involved in this process.
- **Gray Box Testing [58, 60]:** It combines elements of both white and black box testing. the tester has partial knowledge or access to an internal network or web application. the tester starts with user privileges on a host and escalates their privileges to a domain admin. In another way, the tester could obtain access to software code and system architecture diagrams to simulate a scenario where some information is available to an attacker. This approach strikes a balance between reality and in-depth analysis.

6.3.3 Choosing the right team for agile security

Each organization desires to select the appropriate penetration testing team based on different factors including the organization's goals, the desired level of realism, and the specific security challenges presented by agile development cycles. In agile security, collaboration and communication between the testing team and development teams are essential for effective vulnerability identification and mitigation [46]. Moreover, the synergy between penetration testing and ethical hacking is important. Both practices are instrumental in identifying and addressing vulnerabilities, ensuring that organizations can preemptively strengthen the organizations' defenses against potential cyber threats. The effectiveness of defense strategies depends often on the result of dynamic interactions among specialized teams. In the following points, the subsection highlights different facets of security team dynamics, clearing the roles and collaborations that contribute to robust cybersecurity frameworks.

6.3.3.1 Red team vs. blue team dynamics

The interaction between the red team and the blue team is essential as a strategy within the cybersecurity domain [61]. The red team [62], acting as the adversary, pursues to exploit vulnerabilities and expose weaknesses. In contrast, the blue team stands as the organization's defender [63], working to prevent attacks and strengthen the system against potential breaches. In the agile security context, these teams often join forces, offering a comprehensive view of an organization's security conditions. the interaction between both teams will benefit organizations, in four aspects:

1. Identify points of vulnerability and their relation to people, technologies, and systems.
2. Figure out where defensive incident response processes can be enhanced in each step of the attack process.
3. Learn how to detect and prevent a specific type of attack by gaining hands-on experience.
4. Developing plans to respond and fix items to bring the system back to its normal state after an attack.

6.3.3.2 The emergence of the purple team

Purple teams have cyber security professionals who comprehend how to attack and defend systems [64]. This knowledge has typically been obtained through educational industry experiences (Mostly Certified) coming from both the red and blue sides and experience. In other words, it represents a perfect method of cybersecurity that combines all of the offensive and defensive strategies, encapsulated in the concept of the purple team. The purple team approach encourages cooperation between development and security teams, ensuring that security measures seamlessly integrate with agile workflows, thereby enhancing overall system resilience.

Traditionally, red teams simulate cyber-attacks to identify vulnerabilities, while blue teams focus on defending against these simulated attacks. The purple team concept combines these efforts, allowing for a more cooperative and synergistic procedure. This collaboration often occurs during security testing or exercises, where the combined knowledge and skills of offensive and defensive teams contribute to a more comprehensive understanding of an organization's security conditions. In an agile environment, the purple team approach becomes especially valuable, facilitating cooperation between development and security teams [65]. This alignment ensures that security measures are effectively integrated into agile workflows, enhancing the organization's overall resilience against cyber threats. Table 6.1 presents a direct comparison between the red, black, and purple teams.

6.3.3.3 Internal vs. external team dynamics

The internal IT team is the group of on-hand individuals working on-site for the firm or business and serving on-site technology support. On the other hand, an external IT team comprises individuals from outside your organization who are hired to address your IT needs known as a Managed Service Provider (MSP) or IT provider, which brings external expertise to improve the firm technological qualifications.

In cybersecurity, the difference between internal and external teams raises a fine layer of security dynamics. Internal teams, as integral components of the organization, possess insider knowledge required for crafting effective defense strategies. Their familiarity with the inner workings of the company contributes to robust security conditions. Conversely, external teams operate independently, simulating external threats to evaluate and strengthen the organization's security defenses. Recognizing the importance of diverse perspectives, this distinction emphasizes the multifaceted nature of comprehensive cybersecurity strategies.

Table 6.1 Comparison of red team, blue team, and purple team

Aspect	*Red team*	*Blue team*	*Purple team*
Focus	Offensive security	Defensive security	Collaboration between red and blue
Objective	Test and exploit security posture	Protect and defend systems	Improve security effectiveness
Methods	Mimic real-world attackers	Proactive security measures	Joint exercises and simulations
Tools	Penetration testing tools	Security technologies	Combined tools and techniques
Outcome	Identify vulnerabilities	Enhance security posture	Enhance organizational resilience

Internal teams may face challenges if they lack specific skills, requiring investment in training programs or additional hiring. In contrast, collaborating with an external team provides access to a broader range of skills without the need for extensive training. However, it's important to note that specialized services from external teams may come at a premium price. Careful consideration is necessary when weighing the benefits of each approach to ensure an optimal balance between skills, costs, and cybersecurity effectiveness [66].

6.4 PENETRATION TESTING CHALLENGES AND SOLUTIONS IN AGILE SECURITY

6.4.1 Understanding penetration testing

Penetration testing is a proactive strategy for estimating the security of an organization's systems, networks, or applications. It relies on authorized professionals known as penetration testers or ethical hackers who endeavor to exploit vulnerabilities in a controlled environment [39]. The direct objective is to identify weaknesses before malicious actors capitalize on them. Thereby, in the agile security landscape, penetration testing allows organizations to maintain a proactive security stance while accommodating the rapid iterations inherent in agile methodologies, that is done by conducting frequent reviews and employing automated tools.

Ethical hacking shares the overarching purpose of penetration testing, where both concentrate on discovering vulnerabilities through simulated cyber-attacks [39]. Ethical hackers are usually part of a penetration testing team and employ their skills to estimate an organization's security level. Commonly, the term (ethical) emphasizes the authorized and responsible temper of the related activities.

Agile security environments require collaborative efforts with ethical hacking. That means ethical hackers work alongside development teams, and hence both become well-versed in the systems and applications under development. This collaboration promotes communication between security and development professionals, creating a shared responsibility for cybersecurity. Integrating penetration testing and ethical hacking into agile security practices offers several benefits:

- Real-time Vulnerability Identification: In agile development cycles, changes occur rapidly. Penetration testing and ethical hacking enable organizations to identify vulnerabilities in real time, ensuring that security stays an integral part of the development process.
- Collaboration and Communication: Both practices encourage collaboration between security and development teams. Ethical hackers engage with developers, sharing insights and recommendations to address vulnerabilities effectively.
- Continuous Improvement: Agile security adopts the concept of continuous improvement. Penetration testing and ethical hacking contribute to

this by providing continuous reviews, allowing organizations to adjust their security measures to new threats.

- Strategic Risk Mitigation: By proactively identifying and addressing vulnerabilities, penetration testing and ethical hacking assist organizations in strategically mitigating risks. This proactive approach aligns with the agile principle of risk management.

To sum up, the relationship between penetration testing and ethical hacking in agile security is symbiotic [67]. Both practices form a comprehensive strategy for identifying, estimating, and mitigating cybersecurity risks. The collaborative temper of ethical hacking guarantees that security measures align with agile development cycles, enabling dynamic and resilient security statutes [39].

6.4.2 Understanding the hacker intentions

Hackers, in the digital world, are classified into various types according to their motives, methods, and implications [68–70]. On the top, the **White Hat Hackers** [71, 72] represent the virtuous guardians of the cyber world. They are professionals and equipped with expertise in cybersecurity. Also, those hackers operate within the boundaries of authorization and legality. They might be employed by governments or organizations to unearth the systems' vulnerabilities, subsequently strengthening defenses against external threats. Termed ethical hackers mainly work twofold: to uphold the integrity of systems and to help in the endless battle against cyber threats. In contrast to the white hat, the **Black Hat Hackers** [73] represent the malicious intent by breaching systems, stealing data, or wreaking havoc upon unsuspecting targets. Their actions are fueled by personal gain or hatred, where security breaches translate into tangible losses for victims. On the opposite side, the **Red Hat Hackers** [74] are considered the formidable adversaries of the black hat hackers. Like the white hat hackers, the red hat hackers operate their skills to prevent the nefarious schemes of the black hats, entertaining in persistent fights to safeguard systems and neutralize threats.

The **Gray Hat Hackers** [75] exist between the white and black hats, operating without official approval, these hackers blur the lines of ethicality, their intentions oscillating between generosity and self-interest. While their endeavors may not always align with conventional ethical standards, they enjoy the thrill of exploration, probing systems for vulnerabilities with an insatiable curiosity.

Apart from the previous archetypes, the script kiddies or novices in the realm of hacking equipped with ready public scripts and a high desire for discovery are driven more by adolescent impulsivity than genuine malice. Their exploits often spin around disrupting services or websites. Meanwhile, **Green Hat Hackers** [76] represent a more intense search for knowledge, as they steer their attempt at hacking with the aspiration of mastering its complexities and entering the levels of professional hackers. In stark contrast, **Blue Hat Hackers** [77] lack the unselfish tendencies of the green hat, have a hunger for fame, and exploit hacking as a means of ensuring dominance or exacting revenge upon adversaries. Their actions are motivated by personal feuds rather than noble reasons.

In addition to the above, there are different types of actors involved in hacking activities. The first group includes individual hackers who act independently. Then, there are **State or Nation Sponsored Hackers** [78] who are operatives working on behalf of governments to collect intelligence from foreign parts. Another group is called **Hacktivists** [79], who are activists that use hacking as a tool for political or social advocacy. Moreover, there are **Malicious Insiders** [80] (or Whistleblowers), who are individuals working within organizations and who expose confidential information either out of personal grievance or to expose misbehavior.

In today's constantly changing cybersecurity landscape [81], it is becoming increasingly difficult to differentiate between a friend and a competitor as the reasons behind hacking attempts are as varied as the people behind them. As a result, organizations encounter a constant threat of cyber-attacks and must remain awake to protect themselves. In this digital era, the stakes are higher than ever before, and it is crucial to heighten defenses and remain alert against cybercrime [82].

6.4.3 Penetration testing challenges and solutions in agile security

Penetration testing in agile security environments carries onward challenges, necessitating innovative solutions to ensure effective vulnerability estimation and mitigation [9]. While penetration testing is important for identifying vulnerabilities and immunizing defenses, conducting such estimations alongside agile security environments new issues. For instance, agile procedures emphasize rapid iteration and continuous integration but might introduce complexities that traditional penetration testing approaches struggle to address effectively. Below are some of the fundamental challenges encountered in performing penetration testing in agile security, along with corresponding solutions:

- Time Constraints: Agile development cycles are characterized by short iterations, requiring short turnaround times for security estimations. Traditional penetration testing may not align with these timelines, leading to delays in identifying and addressing vulnerabilities. To handle this challenge, the organization, agile development team, or penetration team utilizes automated testing tools [83, 84] (i.e., Nessus [85], OpenVAS [86], Burp Suit [87], Metasploit [88], OWASP ZAP [89], and so on) that seamlessly integrate into the agile development pipeline. Thus, Automation facilitates continuous security testing throughout the development procedure, guaranteeing timely identification and handling of vulnerabilities.
- Collaboration Gaps: Agile development teams usually include cross-functional members performing concurrently and coordinating between

security and development teams' challenges. Communication gaps would result in security considerations being overlooked or poorly addressed. Therefore, fostering a culture of collaboration between security and development teams by promoting regular communication and knowledge sharing. Encourage security professionals to actively participate in agile ceremonies, such as sprint planning and retrospectives, to ensure security concerns are integrated into the development process from the outset.

- Resource Constraints: Limited resources, both in terms of personnel and infrastructure, hinder the effectiveness of penetration testing efforts within agile environments. Security teams would struggle to keep swiftness with the rapid speed of development or lack access to the necessary tools and environments for testing. As a suggested solution, the development teams prioritize resource allocation based on risk assessment and threat modeling. When the penetration testing efforts concentrate on high-priority areas and critical functionalities, that will ensure the optimal benefit of available resources, in addition, to leveraging cloud-based testing environments and scalable infrastructure. As a result, the resource limitations are overcome and the developing and penetration testing teams facilitate comprehensive security testing.
- CI Challenges: Agile development and its CI/CD pipelines can make security testing complex. Integrating security assessments into CI/CD pipelines without interrupting the development workflow is challenging. Therefore, it is crucial to incorporate security testing tools and processes into CI/CD pipelines to enable automated security testing at each stage of the development life cycle. Therefore, it is suggested to set up security gates within the pipeline to trigger alerts or suspend the deployment process when critical security findings are detected. This ensures that vulnerabilities are addressed before production deployment.
- Maintaining Test Coverage: In agile environments, where software applications evolve rapidly, ensuring comprehensive test coverage across all functionalities and components can be challenging. Security testing can struggle to keep pace with the continuous changes and updates introduced during development. To address this issue, the development teams adopt a risk-based approach to penetration testing, focusing on critical areas of the application or infrastructure that pose the highest security risk. Prioritizing testing based on the likelihood and potential impact of exploitation, and allocating resources accordingly, can maximize test coverage where it matters most.

By proactively addressing these challenges and implementing effective solutions, organizations can streamline penetration testing efforts within agile security environments. This enhances their ability to identify and mitigate security vulnerabilities in a timely and efficient manner. Agile development methodologies prioritize rapid iterations and continuous delivery, creating inherent time constraints for penetration testing. Traditional testing models, often lengthy and periodic, struggle to keep pace with the dynamic nature of agile development cycles [90].

In General, some methods for fostering communication and collaboration between development and security teams are crucial for ensuring the success of agile security practices. **Cross-functional training** [91] initiatives present a valuable opportunity for teams to achieve mutual familiarity with their respective objectives, methodologies, and challenges, establishing a shared language for effective communication. Furthermore, **Embedding security champions** [92] within development teams serves as a bridge between the two disciplines, facilitating communication, addressing concerns promptly, and cultivating a culture of security understanding. Moreover, **Joint workshops and agile sprints** [93] applying both development and security teams are instrumental in promoting collaboration and integrating security practices seamlessly into the development process.

Activities such as threat modeling sessions encourage shared problem-solving and reinforce the importance of security considerations. **Utilizing collaboration tools** and platforms improves communication by providing integrated solutions for problem tracking, documentation, and communication. Regular meetings and feedback loops between development and security teams offer opportunities for coordination, collaboration, and endless improvement. All together ensure that security measures develop according to the dynamic development landscape.

6.5 INTEGRATION OF PENETRATION TESTING TOOLS

The main goal of penetration testing is to estimate the security conditions of software applications and systems. Hence, integrating penetration testing tools into CI/CD pipelines is essential for ensuring the security of software applications throughout the development life cycle. Here, the chapter discusses several popular penetration testing tools and how organizations can merge these tools into CI/CD pipelines.

6.5.1 Stages of the integration process into CI/CD pipelines

Integrating penetration testing tools into CI/CD pipelines involves automating their execution at various pipeline stages. This ensures that security is not an afterthought but an integral part of the development process. Here's how it can be achieved in general [94–96]:

1. **Pre-Commit Stage:**
    - Conduct static code analysis using tools like Checkmarx or SonarQube to identify potential security vulnerabilities in the codebase.
    - Run dependency scanning tools like OWASP Dependency-Check to detect known vulnerabilities in third-party libraries.

2. **Commit Stage:**
   - Perform automated security tests using tools like OWASP ZAP or Burp Suite to scan web applications for vulnerabilities.
   - Execute network scans using Nmap to identify potential security issues in the infrastructure.
3. **Build Stage:**
   - Integrate dynamic application security testing (DAST) tools like Metasploit to simulate attacks and exploit known vulnerabilities.
   - Conduct fuzz testing using tools like AFL or Sulley to identify input validation vulnerabilities.
4. **Deployment Stage:**
   - Automate the deployment of security patches or fixes for identified vulnerabilities.
   - Integrate automated retesting to ensure that vulnerabilities have been adequately addressed before deployment.

Integrating penetration testing tools into the CI/CD pipeline enables organizations to make security a constant and essential aspect of the software development life cycle. This proactive approach helps detect and address potential security vulnerabilities at an early stage of the development process, which in turn helps minimize the risks of security breaches in production environments.

6.5.2 Penetration testing tools

1. **Burp Suite** [87]**:**
   - **Functionality:** Burp Suite is a widely used platform for web application security testing to identify vulnerabilities such as SQL injection, cross-site scripting (XSS), and more. It includes tools for manual and automated testing, such as scanning for vulnerabilities, intercepting proxies, and advanced crawling and analysis capabilities.
   - **Integration:** Burp Suite can be integrated into CI/CD pipelines using its REST API. Automated scans can be triggered as part of the pipeline to identify vulnerabilities in web applications during the development process. Integration of Burp Suite into CI/CD Pipelines can be done as the following points:
     a. Custom Plugins: Burp Suite offers an extensible architecture that helps the developers make custom plugins to automate security testing tasks. Developers can leverage these plugins to integrate Burp Suite into CI/CD pipelines.
     b. Headless Mode: Burp Suite supports a headless mode, which automates scanning without the need for a graphical user interface (GUI). This feature is particularly helpful for integrating Burp Suite into automated testing frameworks within CI/CD pipelines.
     c. API Integration: Burp Suite provides a robust REST API that allows developers to automate scanning tasks and retrieve scan

results programmatically. This enables seamless integration with CI/CD pipelines, allowing for automated security testing at every stage of the development life cycle.

d. Continuous Scanning: Integrate Burp Suite into CI/CD pipelines to perform continuous scanning of the application as new code is developed and deployed. This helps identify and address security vulnerabilities promptly.

e. Integration with Issue Trackers: Integrate Burp Suite with issue tracking systems (e.g., JIRA, GitHub Issues) to automatically create tickets for identified vulnerabilities. This streamlines the vulnerability management process and ensures that issues are addressed promptly.

2. **Nmap (Network Mapper) [97]:**
   - **Functionality:** Nmap is a network scanning tool that can discover hosts and services on a computer network. It is well-known for its flexibility, efficiency, and extensive feature set, which make it a popular tool for security professionals.
   - **Integration:** Nmap scans can be integrated into CI/CD pipelines to ensure that network security is tested continuously. For example, before deploying a new service or application, a pre-deployment Nmap scan can be performed to identify any open ports or potential security vulnerabilities. The following key points summarize the integration process into CI/CD pipelines:
     a. Pre-Deployment Scans: Incorporate Nmap scans into the CI pipeline as part of the pre-deployment phase to identify open ports, services, and potential vulnerabilities in the target environment.
     b. Integration with Configuration Management Tools: Integrate Nmap with configuration management tools like Ansible or Puppet to automate network scans across various environments (e.g., development, staging, production) within the CI/CD pipeline.
     c. Scheduled Scans: Schedule periodic Nmap scans to ensure continuous monitoring of network assets and infrastructure. These scans can be triggered automatically as part of the CI/CD pipeline or orchestrated through scheduling tools like corn.
     d. Integration with Vulnerability Management Systems: Integrate Nmap with vulnerability management systems such as OpenVAS or Nessus to correlate scan results with known vulnerabilities, enabling prioritization and remediation efforts.
     e. Custom Scripts and Plugins: Extend Nmap's functionality by developing custom scripts and plugins tailored to the organization's specific requirements. These scripts can automate repetitive tasks, enhance scan accuracy, and integrate seamlessly into CI/CD workflows.
3. **Metasploit [98]:**
   - **Functionality:** Metasploit is a penetration testing framework that enables users to exploit known vulnerabilities in a target system.

It includes a vast array of exploits, payloads, and auxiliary modules, making it a go-to tool for both offensive and defensive security assessments. In other words, Metasploit is a powerful penetration testing framework that facilitates the discovery, exploitation, and validation of security vulnerabilities.

- **Integration:** Metasploit can be integrated into CI/CD pipelines for automated security testing. Vulnerability scans can be conducted using Metasploit's modules to check for common vulnerabilities and exposures (CVEs) in the application or system being developed. Integrating Metasploit into CI/CD pipelines allows for proactive identification and remediation of vulnerabilities before they are deployed to production. the following key points show how organizations can utilize Metasploit in CI/CD pipelines:
  a. Automated Exploitation Testing: Integrate Metasploit into CI/CD pipelines to automate the testing of exploitability for known vulnerabilities identified during development. This ensures that critical vulnerabilities are addressed before deployment.
  b. Validation of Security Controls: Use Metasploit to validate the effectiveness of security controls, such as firewalls and intrusion detection systems, deployed in the target environment. Automated tests can be triggered as part of the CI/CD pipeline to verify the resilience of the infrastructure against common attack vectors.
  c. Integration with Vulnerability Scanners: Combine Metasploit with vulnerability scanners like Nexpose or Qualys to validate and exploit identified vulnerabilities. This holistic approach combines automated scanning with manual validation, providing comprehensive coverage of security assessments.
  d. Continuous Monitoring and Response: Incorporate Metasploit into CI/CD pipelines to monitor and respond to security incidents. Automated scripts can detect and mitigate potential threats in real time, minimizing the impact on the organization's assets and operations.
  e. Custom Exploit Development: Utilize Metasploit's extensible architecture to develop custom exploits and modules tailored to specific applications or environments. These customizations enhance the tool's effectiveness and adaptability within CI/CD workflows.

4. **OWASP ZAP (Zed Attack Proxy) [98]:**
   - **Functionality:** OWASP ZAP is an open-source web application security scanner supported by the Open Web Application Security Project (OWASP). OWASP ZAP is designed to automatically detect vulnerabilities in web applications during the development and testing phases. Also, it is used by both experienced and beginner security professionals.
   - **Integration:** OWASP ZAP can be integrated into CI/CD pipelines to automate security testing of web applications. By incorporating

OWASP ZAP scans into the pipeline, developers can identify and fix security vulnerabilities early in the development process, reducing the risk of deploying insecure code to production.

a. Build Stage Integration: Integrate ZAP into the build process in CI pipelines so that developers can scan the application code base for security vulnerabilities.
b. Automated Scans: Use ZAP's automation capabilities to perform regular scans of the application as new code is added to the repository. This can be achieved by triggering ZAP scans after each code commit or at specific intervals.
c. API Integration: Utilize ZAP's comprehensive API to integrate it with CI/CD pipelines seamlessly. Developers can programmatically trigger ZAP scans and retrieve scan results for further analysis.
d. Thresholds and Policies: Define security thresholds and policies within ZAP to determine which vulnerabilities should trigger a build failure. This ensures that critical vulnerabilities are addressed before deploying the application.
e. Reporting: Generate detailed reports of ZAP scan results in various formats (e.g., HTML, XML) and integrate them into CI/CD pipelines. These reports provide insights into identified vulnerabilities and their severity levels.

Organizations automate identifying and validating security-related issues by integrating security testing tools like OWASP ZAP and Burp Suite into their CI/CD pipelines. This proactive approach helps organizations solve security issues early in the development life cycle, thus reducing the risk of potential breaches. Likewise, integrating network reconnaissance tools like Nmap and vulnerability assessment tools like Metasploit into the CI/CD pipelines.

6.6 NAVIGATING Best PRACTICES AND FUTURE DIRECTION FOR AGILE PENETRATION TESTING

6.6.1 Latest cybersecurity threats and corresponding penetration testing methodologies

In the war between cyber defenders and adversaries, the battleground of cybersecurity always shifts and transforms. With each passing year, new technologies emerge, and with them, novel tactics are invented by cyber-criminals to breach defenses and exploit vulnerabilities. Therefore, understanding the latest advancements in cybersecurity threats is imperative to learn how they influence the landscape of penetration testing methodologies. The following demonstrates the most common cyber threats and their corresponding penetration testing methodologies.

- **Advanced Malware and Ransomware Attacks:** One of the most prevalent cybersecurity threats encountering organizations today is the development of advanced malware and ransomware attacks. These attacks often exploit vulnerabilities in software systems to gain unauthorized access, encrypt data, and request ransom payments. To combat these threats, penetration testing methodologies such as DAST and static application security testing (SAST) are crucial. DAST involves simulating real-world attack scenarios to specify vulnerabilities in running applications. While SAST analyzes source code for potential security vulnerabilities before deployment, providing proactive defense against malware and ransomware attacks [29, 30].
- **APTs:** APTs are highly sophisticated cyberattacks developed by skilled and well-funded threat actors to gain long-term access to sensitive information. These attacks are difficult to detect using traditional penetration testing methods because of their hidden nature and use of advanced evasion techniques. To effectively counter APTs, organizations can use a combination of red team exercises and threat emulation methodologies. Red team involves simulating realistic attack scenarios to thoroughly assess an organization's security status. Threat emulation techniques imitate the tactics, techniques, and procedures operated by APT actors, and help organizations to proactively identify and neutralize potential threats [78].
- **Supply Chain Vulnerabilities:** With the increasingly interconnected digital ecosystems, supply chain vulnerabilities have emerged as a significant cybersecurity problem. Malicious actors often target third-party vendors and suppliers to gain unauthorized access to sensitive data or compromise critical infrastructure. Traditional penetration testing methodologies may supervise supply chain vulnerabilities and emphasize the need for comprehensive supply chain security assessments. These assessments may apply penetration testing of third-party systems and infrastructure, as well as vulnerability assessments and code reviews of software components integrated into the supply chain. Besides, organizations might utilize threat intelligence and information-sharing initiatives to stay up-to-date on emerging supply chain threats and vulnerabilities [99].
- **Insider Threats:** Insider threats represent a major risk to organizations, as malicious insiders or unwitting employees may inadvertently compromise sensitive data or systems. Traditional penetration testing approaches may skip insider threats, as they primarily concentrate on external attack vectors. To address insider threats effectively, organizations must employ social engineering penetration testing methodologies, such as phishing simulations and user awareness training. These activities evaluate employees' vulnerability to social engineering attacks and educate them on best practices for specifying and mitigating insider threats [80].
- **IoT Vulnerabilities:** The widespread of IoT devices presents unique cybersecurity challenges, as these interconnected devices often lack robust security features which make them vulnerable to exploitation by

malicious actors. the existing penetration testing methodologies dedicated to the IoT environments are essential for identifying and mitigating IoT vulnerabilities, such as IoT device testing and network segmentation assessments. IoT device testing works to analyze the security conditions of individual IoT devices, while network segmentation assessments evaluate the effectiveness of network segmentation measures in containing potential IoT-related threats [32, 100, 101].

- **Cloud Security Risks:** Cloud computing improves the method organizations store, process, and manage data. Yet, cloud environments have new security risks, including data breaches, misconfigurations, and unauthorized access. To mitigate cloud security risks, organizations use cloud-specific penetration testing methodologies, such as cloud infrastructure assessments and configuration reviews. These assessments involve evaluating the security controls and configurations of cloud platforms and services to identify and remediate vulnerabilities that could expose sensitive data or compromise cloud infrastructure [31, 102].
- **AI and ML Exploitation:** Although AI and ML technologies become more used in organizational workflows and decision-making processes, they also present new vulnerabilities that might be exploited by malicious actors. Adversarial attacks, data poisoning, and model evasion techniques can compromise the integrity and reliability of AI and ML systems, creating significant cybersecurity risks. To ensure the security of these systems, penetration testing methodologies are used to evaluate the strength and resilience of AI models and algorithms against adversarial attacks. This involves identifying potential vulnerabilities in data pipelines and training processes to control any security breaches [33, 49].

According to the aforementioned cybersecurity threats that become more complex and diverse, organizations have to adopt proactive penetration testing methodologies addressing the specific challenges posed by emerging threat vectors. Also, it's crucial to employ adaptive penetration testing techniques that can keep up with the evolving threat landscape. By integrating proactive security measures and leveraging cutting-edge techniques, organizations will enhance their resilience against emerging cyber threats and protect their digital assets effectively. Thus, organizations can safeguard themselves against the ever-changing cyber threat landscape by staying ahead of the curve and employing adaptive security measures.

6.6.2 Best practices

Through agile penetration testing, certain best practices stand out essentially to ensure the effectiveness and efficiency of security estimations. *Integrating early and often* embeds penetration testing into the early development steps and maintains persistent testing during the agile life cycle. Likewise, *Automating for agility* utilizes automated penetration testing tools to accommodate rapid agile development and ensure instant identification of vulnerabilities. Also, *Prioritizing based on risk* adopts a risk-based technique that allows concentrating testing efforts first on high-impact vulnerabilities. *Cultivating a collaborative culture* between development and security teams enables open communication and shared responsibility for security. Moreover, *Tailoring testing to the environment* is necessary to customize penetration testing techniques, and hence to suit the specific attributes of the organizational environment (whether the IoT, web applications, or other systems) and then enhance the effectiveness.

Once the organization pursues to implement these best practices effectively, several actionable steps should be considered. In the beginning, *Investing in training* for both development and security teams will enhance the team members' knowledge of agile security practices and maximize their creativity and accuracy. Similarly, *Establishing clear processes* combines penetration testing into agile workflows and ensures that security measures compatible with development activities efficiently. On the other side, *Regularly reviewing and updating* penetration testing strategies are necessary to adapt to evolving threats and changes in the agile environment. Lastly, *Celebrating successes and learning from failures* promotes a culture of continuous improvement, celebrating successful implementations and learning from any challenges or failures experienced during the penetration testing process.

In adopting these best practices and drawing inspiration from successful case studies, organizations strengthen their agile security strategies, delivering secure software products in an agile and responsive manner [46, 103–105].

6.6.3 Case studies: Agile penetration testing across industries

Cybersecurity challenges are a widespread issue across various industries in today's rapidly changing digital landscape. To tackle these challenges effectively, many organizations have begun adopting agile penetration testing frameworks to support their specific environments and requirements. Hence, it is essential to understand a range of case studies from different industries to demonstrate how the agile penetration testing approach can be adapted and implemented successfully. The following case studies show that agile penetration testing is an adaptable and effective technique for cybersecurity across various industries and environments.

1. **Healthcare:** A healthcare organization encountered the challenge of protecting patient health information across its diverse ecosystem of medical devices, electronic health records (EHR) systems, and third-party applications [106]. To address this issue, the organization adopted an agile penetration testing framework, which allowed it to continuously test its systems and applications and ensure compliance with healthcare regulations such as the Health Insurance Portability and Accountability Act (HIPAA) [107]. By using this agile approach, the organization could

proactively identify and address security vulnerabilities, thus safeguarding patient data and maintaining the integrity of its healthcare services.

2. **E-commerce:** An e-commerce company operating a large-scale online platform sought to protect customer data and secure its payment processing systems from cyber threats. By integrating agile penetration testing into its development pipeline, the company conducted regular security assessments of its web applications, APIs, and payment gateways. This proactive approach enabled the company to identify and mitigate vulnerabilities quickly, maintaining customer trust and confidence in its online platform.
3. **Financial Services:** A multinational bank has implemented an agile penetration testing framework to improve its cybersecurity. This framework involves frequent and targeted tests that are aligned with the bank's agile development cycles. By doing so, the bank can quickly identify and fix vulnerabilities in critical systems. This approach also helps the bank remain compliant with regulatory requirements and effectively protect sensitive financial data from cyber threats.
4. **Technology**: A technology firm specializing in software development faced the challenge of securing its cloud-based infrastructure and DevOps tools used for CI and CD. Leveraging an agile penetration testing framework, the firm conducted automated security testing throughout the software development life cycle, from code commits to production releases. By integrating security into its agile development processes, the firm minimized the risk of security breaches and ensured the reliability and security of its software products.
5. **Government:** A government agency responsible for critical infrastructure protection implemented an agile penetration testing framework to assess the security of its network infrastructure and ICS. By conducting regular penetration tests aligned with agile sprints, the agency identified vulnerabilities in its ICS environments and implemented targeted correction measures to mitigate potential cyber threats. This agile approach enhanced the resilience of the agency's critical infrastructure against evolving cybersecurity risks.

6.7 FUTURE TRENDS AND RECAP

6.7.1 Future research directions

In agile security, as long as emerging cyber threats and advances in technology, penetration testing is continually developing in response. For that, utilizing AI and ML techniques in penetration testing methodologies is a promising area of research [108–110]. AI and ML will enrich the testing process by automating repetitive tasks, identifying patterns in system behavior, and detecting anomalies that may indicate potential vulnerabilities. Thereby, when the organization leverages AI-powered tools for threat detection [102, 111], the penetration testing process will be facilitated for faster identification and treatment of security shortcomings. Correspondingly, blockchain technology will offer secure platforms for storing and sharing sensitive testing data [112, 113]. Hence, it will improve the security and integrity of penetration testing processes. Research areas to investigate novel applications of blockchain technology would be another promising method to improve the reliability and confidentiality of penetration testing operations. The expansion of IoT devices and connected systems represents a big challenge for penetration testing in agile security [100]. Therefore, future research directions would focus on developing technological testing methodologies and tools working due to the features of IoT environments. Besides, improvements in edge computing and distributed systems require clever methods of security testing compatible with the decentralized nature of these infrastructures. To sum up, the developers will at most enhance the security conditions of IoT deployments and mitigate the risks associated with interconnected devices by including IoT-based testing strategies and advanced analytics [101].

Other areas of interest include quantum computing [114], augmented and virtual reality [115], containerization [116], microservices architectures [117], and zero-trust security architectures [118]. Researchers should investigate these emerging technologies and integrate them into penetration testing procedures to continuously verify and implement security policies. Moving forward, as developers stay up-to-date with these advancements, they will expand their range of penetration testing techniques, thereby enhancing their readiness to tackle the ever-evolving cybersecurity challenges. Figure 6.3 provides a concise overview of several potential areas for future research.

Regarding the outlook work, there are numerous possibilities for further exploration. A promising direction lies in refining penetration testing techniques specifically tailored for Agile environments. By leveraging automation and cutting-edge algorithms, we can revolutionize the efficiency and accuracy of penetration testing, especially in the high-speed environments typical of Agile projects. Additionally, organizations must understand the interplay between regulatory requirements and Agile methodologies to maintain compliance standards while sustaining agility and innovation.

However, we must acknowledge the limitations of our study. We mainly focused on theoretical frameworks and foundational concepts, leaving ample room for empirical investigations in real-world settings. Furthermore, the ever-evolving nature of Agile methodologies and cybersecurity landscapes requires ongoing adaptation and vigilance. What holds today may become obsolete tomorrow, emphasizing the need for continuous research and flexibility in our approaches. Additionally, the applicability of our study may be limited by the specific organizational contexts examined, highlighting the importance of embracing diverse perspectives to gain a comprehensive understanding of agile security and penetration testing.

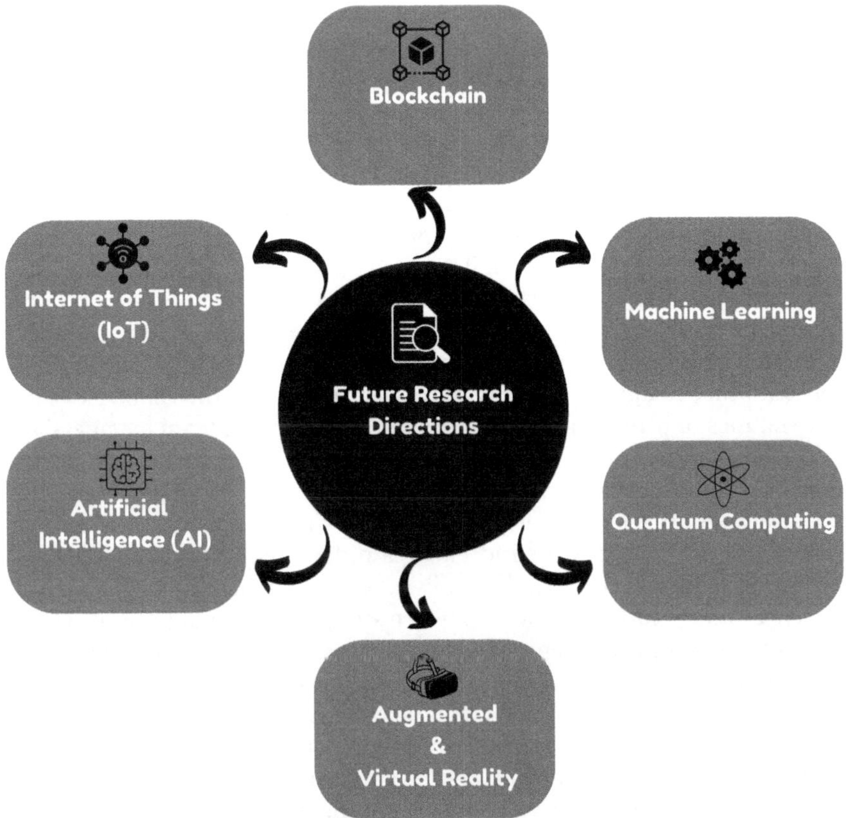

Figure 6.3 Potential future research directions.

6.8 CONCLUSION

The future of penetration testing is related to the evolving landscape of agile security practices. Agile methodologies highlight CI and delivery and require a smooth integration of security into the development life cycle. The adoption of DevSecOps is an essential direction to prioritize the integration of security practices with the software development process. This shift requires organizations to adopt more automated and integrated penetration testing methods within agile environments to quickly identify and handle vulnerabilities. Additionally, future penetration testing in agile security will focus more on threat modeling and risk assessment, enabling efficient resource allocation and proactive vulnerability management.

To help organizations steer challenges related to time constraints, collaboration, and resource management, the security and development teams explore various aspects of penetration testing within agile environments and can cultivate a proactive security culture that adapts to the dynamic nature of agile development by following best practices for agile penetration testing. Embracing automated testing tools, risk-based prioritization, cross-functional training, and continuous collaboration is essential for safeguarding against vulnerabilities. In the end, integrating penetration testing into agile security practices is a strategic imperative that represents an ongoing commitment to continuous improvement, collaboration, and adaptability in the face of evolving cyber threats.

In this chapter, a comprehensive exploration is provided for the integration of penetration testing into an agile security environment. In the beginning, the chapter introduces an overview of agile security principles which emphasizes the essential role of penetration testing in mitigating security threats by examining the types, components, and steps of penetration testing, clarifying their relevance to agile security practices. Furthermore, the chapter discusses the dynamics of penetration testing teams, summarizing team members' roles and responsibilities and addressing the challenges inherent in agile environments. Moreover, the chapter shows the classification of penetration testing and insights into selecting the right team for agile security, including considerations such as red team and blue team dynamics and the emergence of the purple team concept. Additionally, the chapter highlights the common challenges and solutions in penetration testing within an agile security environment. As well, it also emphasizes the importance of facilitating communication and collaboration among team members to ensure effective security testing outcomes. Finally, the chapter steers best practices and future directions for agile penetration testing, offering recommendations for optimizing security testing processes and suggesting areas for further research. In conclusion, it highlights the significance of integrating penetration testing into agile security practices to enhance cybersecurity resilience.

REFERENCES

1. Beznosov, K., Kruchten, P.: Towards agile security assurance. In: Proceedings of the 2004 Workshop on New Security Paradigms. NSPW '04, pp. 47–54. Association for Computing Machinery, New York, NY, USA (2004). https://doi.org/10.1145/1065907.1066034
2. Rindell, K., Ruohonen, J., Holvitie, J., Hyrynsalmi, S., Lepp¨anen, V.: Security in agile software development: A practitioner survey. Information and Software Technology **131**, 106488 (2021).
3. Kongsli, V.: Towards agile security in web applications. In: Companion to the 21st ACM SIGPLAN Symposium on Object-Oriented Programming Systems, Languages, and Applications. OOPSLA '06, pp. 805–808. Association for Computing Machinery, New York, NY (2006). https://doi.org/10.1145/1176617.1176727
4. Bezerra, C.M.M., Sampaio, S.C., Marinho, M.L.: Secure agile software development: policies and practices for agile teams. In: International Conference on the Quality of Information and Communications Technology, pp. 343–357 (2020). Springer Nature Switzerland AG. https://link.springer.com/chapter/10.1007/978-3-030-58793-2_28#citeas

5. Dove, R., Shirey, L.: On discovery and display of agile security patterns. In: Conference on Systems Engineering Research, Stevens Institute of Technology, Hoboken, NJ (2010)
6. Keramati, H., Mirian-Hosseinabadi, S.-H.: Integrating software development security activities with agile methodologies. In: 2008 IEEE/ACS International Conference on Computer Systems and Applications, pp. 749–754 (2008). https://doi.org/10.1109/AICCSA.2008.4493611
7. Kongsli, V.: Towards agile security in web applications. In: Companion to the 21st ACM SIGPLAN Symposium on Object-oriented Programming Systems, Languages, and Applications, pp. 805–808 (2006).
8. Arkin, B., Stender, S., McGraw, G.: Software penetration testing. IEEE Security & Privacy **3**(1), 84–87 (2005).
9. Tomanek, M., Klima, T.: Penetration testing in agile software development projects. arXiv preprint arXiv:1504.00942 (2015).
10. Hamisi, N.Y., Mvungi, N.H., Mfinanga, D.A., Mwinyiwiwa, B.M.: Intrussion detection by penetration test in an organization network. In: 2009 2nd International Conference on Adaptive Science & Technology (ICAST), pp. 226–231 (2009). IEEE
11. Franqueira, V.N., Bakalova, Z., Tun, T.T., Daneva, M.: Towards agile security risk management in re and beyond. In: Workshop on Empirical Requirements Engineering (EmpiRE 2011), pp. 33–36 (2011). IEEE
12. Bou Ghantous, G., Gill, A.: DevOps: Concepts, practices, tools, benefits and challenges. PACIS2017 (2017).
13. Arachchi, S., Perera, I.: Continuous integration and continuous delivery pipeline automation for agile software project management. In: 2018 Moratuwa Engineering Research Conference (MERCon), pp. 156–161 (2018). IEEE.
14. Mansfield-Devine, S.: DevOps: Finding room for security. Network Security **2018**(7), 15–20 (2018).
15. Bin Arfaj, B.A., Mishra, S., Alshehri, M.: Efficacy of unconventional penetration testing practices. Intelligent Automation & Soft Computing **31**(1) (2022).
16. Muhammad, T., Munir, M.T., Munir, M.Z., Zafar, M.W.: Integrative cybersecurity: Merging zero trust, layered defense, and global standards for a resilient digital future. International Journal of Computer Science and Technology **6**(4), 99–135 (2022).
17. Blancaflor, E., Billo, H.K.S., Saunar, B.Y.P., Dignadice, J.M.P., Domondon, P.T.: Penetration assessment and ways to combat attack on android devices through stormbreaker—A social engineering tool. In: 2023 6th International Conference on Information and Computer Technologies (ICICT), pp. 220–225 (2023). IEEE.
18. Shinde, P.S., Ardhapurkar, S.B.: Cyber security analysis using vulnerability assessment and penetration testing. In: 2016 World Conference on Futuristic Trends in Research and Innovation for Social Welfare (Startup Conclave), pp. 1–5 (2016). IEEE.
19. Spremi´c, M., Šimunic, A.: Cyber security challenges in digital economy. In: Proceedings of the World Congress on Engineering **1**, 341–346 (2018). International Association of Engineers Hong Kong, China
20. Sfar, A.R., Natalizio, E., Challal, Y., Chtourou, Z.: A roadmap for security challenges in the internet of things. Digital Communications and Networks **4**(2), 118–137 (2018).
21. Tounsi, W., Rais, H.: A survey on technical threat intelligence in the age of sophisticated cyber attacks. Computers & Security **72**, 212–233 (2018).
22. Berg, K., Crawford, J.N., Seymour, T.: Unbreakable: A concise overview of cybersecurity. Issues in Information Systems **17**(4) (2016)
23. Roshanaei, M.: Enhancing mobile security through comprehensive penetration testing. Journal of Information Security **15**(2), 63–86 (2024).

24. Salahdine, F., Kaabouch, N.: Social engineering attacks: A survey. Future Internet **11**(4), 89 (2019).
25. Ivaturi, K., Janczewski, L.: A taxonomy for social engineering attacks (2011). https://aisel.aisnet.org/cgi/viewcontent.cgi?article=1015&context=confirm2011
26. Al-Otaibi, A.F., Alsuwat, E.S.: A study on social engineering attacks: Phishing attack. International Journal of Recent Advances in Multidisciplinary Research 7(11), 6374–6380 (2020).
27. Workman, M.: Wisecrackers: A theory-grounded investigation of phishing and pretext social engineering threats to information security. Journal of the American Society for Information Science and Technology **59**(4), 662–674 (2008).
28. Hadnagy, C.: Social Engineering: The Art of Human Hacking. John Wiley & Sons (2010). https://www.wiley.com/en-us/Social+Engineering%3A+The+Art+of+Human+Hacking-p-9780470639535
29. Mohurle, S., Patil, M.: A brief study of wannacry threat: Ransomware attack 2017. International Journal of Advanced Research in Computer Science **8**(5), 1938–1940 (2017).
30. Kharraz, A., Robertson, W., Balzarotti, D., Bilge, L., Kirda, E.: Cutting the Gordian knot: A look under the hood of ransomware attacks. In: Detection of Intrusions and Malware, and Vulnerability Assessment: 12th International Conference, DIMVA 2015, Milan, Italy, July 9-10, 2015, Proceedings 12, pp. 3–24 (2015). Springer
31. Kandukuri, B.R., Rakshit, A., *et al.*: Cloud security issues. In: 2009 IEEE International Conference on Services Computing, pp. 517–520 (2009). IEEE.
32. Mohanta, B.K., Jena, D., Satapathy, U., Patnaik, S.: Survey on IoT security: Challenges and solution using machine learning, artificial intelligence and blockchain technology. Internet of Things **11**, 100227 (2020).
33. Li, J.-h: Cyber security meets artificial intelligence: A survey. Frontiers of Information Technology & Electronic Engineering **19**(12), 1462–1474 (2018).
34. Hagos, D.H., Yazidi, A., Kure, Ø, Engelstad, P.E.: Enhancing security attacks analysis using regularized machine learning techniques. In: 2017 IEEE 31st International Conference on Advanced Information Networking and Applications (AINA), pp. 909–918 (2017). IEEE.
35. Kaloudi, N., Li, J.: The AI-based cyber threat landscape: A survey. ACM Computing Surveys (CSUR) **53**(1), 1–34 (2020).
36. Tankard, C.: What the GDPR means for businesses. Network Security **2016**(6), 5–8 (2016).
37. Harding, E.L., Vanto, J.J., Clark, R., Hannah Ji, L., Ainsworth, S.C.: Understanding the scope and impact of the California consumer privacy act of 2018. Journal of Data Protection & Privacy **2**(3), 234–253 (2019).
38. Bacudio, A.G., Yuan, X., Chu, B.-T.B., Jones, M.: An overview of penetration testing. International Journal of Network Security & Its Applications **3**(6), 19 (2011).
39. Engebretson, P.: The Basics of Hacking and Penetration Testing: Ethical Hacking and Penetration Testing Made Easy. Elsevier (2013). https://books.google.de/books?hl=en&lr=&id=69dEUBJKMiYC&oi=fnd&pg=PP1&dq=The+Basics+of+Hacking+and+Penetration+Testing:+Ethical+Hacking+and+Penetration+Testing+Made+Easy.+Elsevier+(2013).&ots=uYP7J6yfFv&sig=oKGuPfZ-HgmtdpwwNpEug4HaJqQ&redir_esc=y#v=onepage&q=The%20Basics%20of%20Hacking%20and%20Penetration%20Testing%3A%20Ethical%20Hacking%20and%20Penetration%20Testing%20Made%20Easy.%20Elsevier%20(2013).&f=false
40. Denis, M., Zena, C., Hayajneh, T.: Penetration testing: Concepts, attack methods, and defense strategies. In: 2016 IEEE Long Island Systems, Applications and Technology Conference (LISAT), pp. 1–6 (2016). IEEE
41. Altulaihan, E.A., Alismail, A., Frikha, M.: A survey on web application penetration testing. Electronics **12**(5), 1229 (2023).

42. Haq, I.U., Khan, T.A.: Penetration frameworks and development issues in secure mobile application development: A systematic literature review. IEEE Access **9**, 87806–87825 (2021).
43. Singh, H., Singh, J.: Penetration testing in wireless networks. International Journal of Advanced Research in Computer Science **8**(5) (2017).
44. Fatima, A., Khan, T.A., Abdellatif, T.M., Zulfiqar, S., Asif, M., Safi, W., Al Hamadi, H., Al-Kassem, A.H.: Impact and research challenges of penetrating testing and vulnerability assessment on network threat. In: 2023 International Conference on Business Analytics for Technology and Security (ICBATS), pp. 1–8 (2023). IEEE.
45. Deng, G., Liu, Y., Mayoral-Vilches, V., Liu, P., Li, Y., Xu, Y., Zhang, T., Liu, Y., Pinzger, M., Rass, S.: Pentestgpt: An llm-empowered automatic penetration testing tool. arXiv preprint arXiv:2308.06782 (2023)
46. Cruzes, D.S., Felderer, M., Oyetoyan, T.D., Gander, M., Pekaric, I.: How is security testing done in agile teams? a cross-case analysis of four software teams. In: Agile Processes in Software Engineering and Extreme Programming: 18th International Conference, XP 2017, Cologne, Germany, May 22-26, 2017, Proceedings 18, pp. 201–216 (2017). Springer International Publishing
47. Sharma, A., Bawa, R.: Identification and integration of security activities for secure agile development. International Journal of Information Technology **14**(2), 1117–1130 (2022).
48. Formosa, P., Wilson, M., Richards, D.: A principlist framework for cybersecurity ethics. Computers & Security **109**, 102382 (2021).
49. Santosh, K., Gaur, L., Santosh, K., Gaur, L.: Privacy, security, and ethical issues. Artificial Intelligence and Machine Learning in Public Healthcare: Opportunities and Societal Impact, 65–74 (2021). https://books.google.de/books?hl=en&lr=&id=bWRXEAAAQBAJ&oi=fnd&pg=PR7&dq=Artificial+Intelligence+and+Machine+Learning+in+Public+Healthcare:+Opportunities+and+Societal+Impact,&ots=2IFrTHzwWt&sig=ILFB9591_6lq07nemGsBLFBCOg8&redir_esc=y#v=onepage&q=Artificial%20Intelligence%20and%20Machine%20Learning%20in%20Public%20Healthcare%3A%20Opportunities%20and%20Societal%20Impact%2C&f=false
50. Hummel, P., Braun, M., Dabrock, P.: Own data? Ethical reflections on data ownership. Philosophy & Technology **34**(3), 545–572 (2021).
51. Mota, G.: The consent for the intervention of the social worker in the protection of fundamental rights within the scope of social insertion income. The Social World after COVID-19 **96** (2022).
52. Rosenberg-Kima, R.B., Koren, Y., Gordon, G.: Robot-supported collaborative learning (RSCL): Social robots as teaching assistants for higher education small group facilitation. Frontiers in Robotics and AI **6**, 148 (2020).
53. Tudosi, A.-D., Graur, A., Balan, D.G., Potorac, A.D.: Research on security weakness using penetration testing in a distributed firewall. Sensors **23**(5), 2683 (2023).
54. Maier, P., Ma, Z., Bloem, R.: Towards a secure scrum process for agile web application development. In: Proceedings of the 12th International Conference on Availability, Reliability and Security, pp. 1–8 (2017)
55. Gutfleisch, M., Schöps, M., Horstmann, S.A., Wichmann, D., Sasse, M.A.: Security champions without support: Results from a case study with OWASP SAMM in a large-scale e-commerce enterprise. In: Proceedings of the 2023 European Symposium on Usable Security, pp. 260–276 (2023)
56. Green, A.: The many faces of lifelong learning: Recent education policy trends in Europe. Journal of Education Policy **17**(6), 611–626 (2002).
57. Ostrand, T.: White-box testing. Encyclopedia of Software Engineering (2002). https://dl.acm.org/doi/abs/10.5555/515664

58. Khan, M.E., Khan, F.: A comparative study of white box, black box and grey box testing techniques. International Journal of Advanced Computer Science and Applications **3**(6) (2012).
59. Beizer, B.: Black-Box Testing: Techniques for Functional Testing of Software and Systems. John Wiley & Sons, Inc. (1995). https://dl.acm.org/doi/abs/10.5555/202699
60. Coulter, A.C.: Graybox software testing methodology: embedded software testng technique. In: Gateway to the New Millennium. 18th Digital Avionics Systems Conference. Proceedings (Cat. No. 99CH37033), vol. 2, p. 10 (1999). IEEE
61. Mejia, R.: Red team versus blue team: How to run an effective simulation. CSO Online-Security and Risk (2008).
62. Mansfield-Devine, S.: The best form of defence—The benefits of red teaming. Computer Fraud & Security **2018**(10), 8–12 (2018).
63. Sehgal, K., Thymianis, N.: Cybersecurity Blue Team Strategies: Uncover the Secrets of Blue Teams to Combat Cyber Threats in Your Organization. Packt Publishing Ltd. (2023). https://books.google.de/books?hl=en&lr=&id=ojWxEAAAQBAJ&oi=fnd&pg=PP1&dq=Cybersecurity+Blue+Team+Strategies:+Uncover+the+Secrets+of+Blue+Teams+to+Combat+Cyber+Threats+in+Your+Organization&ots=5W3xnzL79m&sig=ifsjkOzOD4hMbVtB9ln_eIf2Kzc&redir_esc=y#v=onepage&q=Cybersecurity%20Blue%20Team%20Strategies%3A%20Uncover%20the%20Secrets%20of%20Blue%20Teams%20to%20Combat%20Cyber%20Threats%20in%20Your%20Organization&f=false
64. Van Buggenhout, E.: Purple teaming: A comprehensive and collaborative approach to cyber security. Cyber Security: A Peer-Reviewed Journal **7**(3), 207–216 (2024).
65. Chowdhury, S.S.: Perceptions of purple teams among cybersecurity professionals. PhD thesis, Purdue University (2019).
66. Zabicki, R., Ellis, S.R.: Penetration testing. In: Computer and Information Security Handbook, pp. 1031–1038. Elsevier (2017). https://www.sciencedirect.com/science/article/abs/pii/B9780128038437000752
67. Baloch, R.: Ethical Hacking and Penetration Testing Guide. Auerbach Publications (2017). https://www.taylorfrancis.com/books/mono/10.4324/9781315145891/ethical-hacking-penetration-testing-guide-rafay-baloch
68. Chng, S., Lu, H.Y., Kumar, A., Yau, D.: Hacker types, motivations and strategies: A comprehensive framework. Computers in Human Behavior Reports **5**, 100167 (2022).
69. Hald, S.L., Pedersen, J.M.: An updated taxonomy for characterizing hackers according to their threat properties. In: 2012 14th International Conference on Advanced Communication Technology (ICACT), pp. 81–86 (2012). IEEE.
70. Goerzen, M., Coleman, G.: Wearing many hats. Data and Society (2022).
71. Jagnarine, A.A.: The role of white hat hackers in information security (2005).
72. Caldwell, T.: Ethical hackers: Putting on the white hat. Network Security **2011**(7), 10–13 (2011).
73. Silic, M., Lowry, P.B.: Breaking bad in cyberspace: Understanding why and how black hat hackers manage their nerves to commit their virtual crimes. Information Systems Frontiers **23**, 329–341 (2021).
74. Filiol, E., Mercaldo, F., Santone, A.: A method for automatic penetration testing and mitigation: A red hat approach. Procedia Computer Science **192**, 2039–2046 (2021).
75. Kirsch, C.: The grey hat hacker: Reconciling cyberspace reality and the law. N. Ky. L. Rev **41**, 383 (2014).
76. Vishnuram, G., Tripathi, K., Tyagi, A.K.: Ethical hacking: Importance, controversies and scope in the future. In: 2022 International Conference on Computer Communication and Informatics (ICCCI), pp. 01–06 (2022). IEEE.
77. Sahu, P.K., Acharya, B.: A review paper on ethical hacking. International Journal of Advanced Research in Engineering and Technology **11**(12), 163–168 (2020).

78. Vitkowsky, V.J.: War Exclusions and Cyber Threats from States and State-Sponsored Hackers. Seiger Gfeller Laurie, LLP, New York (2017).
79. Taylor, P.A.: From hackers to hacktivists: Speed bumps on the global superhighway? New Media & Society 7(5), 625–646 (2005).
80. Maybury, M., Chase, P., Cheikes, B., Brackney, D., Matzner, S., Hetherington, T., Wood, B., Sibley, C., Marin, J., Longstaff, T., *et al.*: Analysis and detection of malicious insiders. In: International Conference on Intelligence Analysis, vol. 5 (2005).
81. Kumar, M., Darshan, S.S., Yarlagadda, V., *et al.*: Introduction to the cybersecurity landscape. In: Malware Analysis and Intrusion Detection in Cyber-Physical Systems, pp. 1–21. IGI Global (2023). https://www.igi-global.com/chapter/introduction-to-the-cyber-security-landscape/331297
82. Bendovschi, A.: Cyber-attacks–trends, patterns and security countermeasures. Procedia Economics and Finance **28**, 24–31 (2015).
83. Shah, M.P.: Comparative analysis of the automated penetration testing tools. PhD thesis, Dublin, National College of Ireland (2020).
84. Al Shebli, H.M.Z., Beheshti, B.D.: A study on penetration testing process and tools. In: 2018 IEEE Long Island Systems, Applications and Technology Conference (LISAT), pp. 1–7 (2018). IEEE.
85. Kumar, H.: Learning Nessus for Penetration Testing. Packt Publishing (2014). https://github.com/InspectorDidi/Hacking-Books/blob/master/Learning%20Nessus%20for%20Penetration%20Testing.pdf
86. Rahalkar, S., Rahalkar, K.: Quick Start Guide to Penetration Testing. Springer, Berkeley, CA (2019).
87. Wear, S.: Burp Suite Cookbook: Practical Recipes to Help You Master Web Penetration Testing with Burp Suite. Packt Publishing Ltd (2018). https://digtvbg.com/files/books-for-hacking/Burp%20Suite%20Cookbook%20-%20Practical%20recipes%20to%20help%20you%20master%20web%20penetration%20testing%20with%20Burp%20Suite%20by%20Sunny%20Wear.pdf
88. Kennedy, D., O'gorman, J., Kearns, D., Aharoni, M.: Metasploit: The Penetration Tester's Guide. Publisher: William Pollock, No Starch Press (2011). https://www.kea.nu/files/textbooks/humblesec/metasploit_apenetrationtestersguide.pdf
89. Nagpure, S., Kurkure, S.: Vulnerability assessment and penetration testing of web application. In: 2017 International Conference on Computing, Communication, Control and Automation (ICCUBEA), pp. 1–6 (2017). IEEE.
90. Rehman, A.U., Nawaz, A., Ali, M.T., Abbas, M.: A comparative study of agile methods, testing challenges, solutions & tool support. In: 2020 14th International Conference on Open Source Systems and Technologies (ICOSST), pp. 1–5 (2020). IEEE.
91. Chau, T., Maurer, F.: Knowledge sharing in agile software teams. In: Logic Versus Approximation: Essays Dedicated to Michael M. Richter on the Occasion of His 65th Birthday, pp. 173–183. Springer.
92. Kim, G.S., Kim, J., Kim, J.: Embedding security into DevOps: Investigating DevSecOps in government software development environment (2021).
93. Paasivaara, M., Durasiewicz, S., Lassenius, C.: Distributed agile development: Using scrum in a large project. In: 2008 IEEE International Conference on Global Software Engineering, pp. 87–95 (2008). IEEE.
94. Abiola, O.B., Olufemi, O.G.: An enhanced CICD pipeline: A DevSecOps approach. International Journal of Computer Applications **975**, 8887 (2023).
95. Padhye, P., Khawale, A., Ansari, G.A.: Streamlining software development: An approach to ci/cd pipeline automation. Vidhyayana-An International Multidisciplinary Peer-Reviewed E-Journal-ISSN 2454-8596 **8**(si7), 505–519 (2023).

96. Kromer, M.: Basics of ci/cd and pipeline scheduling. In: Mapping Data Flows in Azure Data Factory: Building Scalable ETL Projects in the Microsoft Cloud, pp. 139–154. Springer (2022). https://link.springer.com/book/10.1007/978-1-4842-8612-8
97. Kaur, G., Kaur, N.: Penetration testing–reconnaissance with Nmap tool. International Journal of Advanced Research in Computer Science **8**(3), 844–846 (2017).
98. Krishnama, S., *et al.*: A process of penetration testing using various tools. Mesopotamian Journal of CyberSecurity **2023**, 93–103 (2023)
99. Sharma, S.K., Srivastava, P.R., Kumar, A., Jindal, A., Gupta, S.: Supply chain vulnerability assessment for manufacturing industry. Annals of Operations Research **326**(2), 653–683 (2023).
100. Yaacoub, J.-P.A., Noura, H.N., Salman, O., Chehab, A.: Ethical hacking for IoT: Security issues, challenges, solutions and recommendations. Internet of Things and Cyber-Physical Systems **3**, 280–308 (2023).
101. Greco, C., Fortino, G., Crispo, B., Choo, K.-K.R.: Ai-enabled IoT penetration testing: State-of-the-art and research challenges. Enterprise Information Systems **17**(9), 2130014 (2023).
102. Hussain, A., and Shabir, G. "AI-Powered DevSecOps: Elevating Security Practices with Machine Learning." https://www.researchgate.net/profile/Ghulam-Shabir-18/publication/380270996_AI-Powered_DevSecOps_Elevating_Security_Practices_with_Machine_Learning/links/66338ee706ea3d0b741fb34f/AI-Powered-DevSecOps-Elevating-Security-Practices-with-Machine-Learning.pdf
103. Türpe, S., Kocksch, L., Poller, A.: Penetration tests a turning point in security practices? organizational challenges and implications in a software development {Team}. In: Twelfth Symposium on Usable Privacy and Security (SOUPS 2016) (2016).
104. Erdogan, G., Meland, P.H., Mathieson, D.: Security testing in agile web application development-a case study using the east methodology. In: Agile Processes in Software Engineering and Extreme Programming: 11th International Conference, XP 2010, Trondheim, Norway, June 1-4, 2010. Proceedings 11, pp. 14–27 (2010). Springer
105. Chòliz, J., Vilas, J., Moreira, J.: Independent security testing on agile software development: a case study in a software company. In: 2015 10th International Conference on Availability, Reliability and Security, pp. 522–531 (2015). IEEE
106. Shah, V., Thakkar, V., Khang, A.: Electronic health records security and privacy enhancement using blockchain technology. In: Data-Centric AI Solutions and Emerging Technologies in the Healthcare Ecosystem, pp. 1–13. CRC Press (2023). https://doi.org/10.1201/9781003356189
107. Moore, W., Frye, S.: Review of HIPAA, part 2: Limitations, rights, violations, and role for the imaging technologist. Journal of Nuclear Medicine Technology **48**(1), 17–23 (2020).
108. Confido, A., Ntagiou, E.V., Wallum, M.: Reinforcing penetration testing using ai. In: 2022 IEEE Aerospace Conference (AERO), pp. 1–15 (2022). IEEE
109. Chaudhary, P.K.: Ai, ml, and large language models in cybersecurity. Volume:06/Issue:02/February-2024.
110. Temara, S.: Maximizing penetration testing success with effective reconnaissance techniques using ChatGPT. arXiv preprint arXiv:2307.06391 (2023).
111. Pulyala, S.R.: From detection to prediction: Ai-powered SIEM for proactive threat hunting and risk mitigation. Turkish Journal of Computer and Mathematics Education (TURCOMAT) **15**(1), 34–43 (2024).
112. Han, H., Shiwakoti, R.K., Jarvis, R., Mordi, C., Botchie, D.: Accounting and auditing with blockchain technology and artificial intelligence: A literature review. International Journal of Accounting Information Systems **48**, 100598 (2023).

113. Gomez Ramirez, A., Al Sardy, L., Gomez Ramirez, F.: Tikuna: An Ethereum blockchain network security monitoring system. In: International Conference on Information Security Practice and Experience, pp. 462–476 (2023). Springer.
114. Brijwani, G.N., Ajmire, P.E., Thawani, P.V.: Future of quantum computing in cyber security. In: Handbook of Research on Quantum Computing for Smart Environments, pp. 267–298. IGI Global (2023). DOI: 10.4018/978-1-6684-6697-1.ch016
115. Oyewole, O.O., Fakeyede, O.G., Okeleke, E.C., Apeh, A.J., Adaramodu, O.R.: Security considerations and guidelines for augmented reality implementation in corporate environments. Computer Science & IT Research Journal **4**(2), 69–84 (2023).
116. VS, D.P., Sethuraman, S.C., Khan, M.K.: Container security: Precaution levels, mitigation strategies, and research perspectives. Computers & Security, 135, 103490 (2023).
117. Nasab, A.R., Shahin, M., Raviz, S.A.H., Liang, P., Mashmool, A., Lenarduzzi, V.: An empirical study of security practices for microservices systems. Journal of Systems and Software **198**, 111563 (2023).
118. Phiayura, P., Teerakanok, S.: A comprehensive framework for migrating to zero trust architecture. IEEE Access **11**, 19487–19511 (2023).

Chapter 7

Forecasting Moroccan stock market using deep learning approaches

Walid El Afari, Sokaina El Khamlichi, Sanaa El Mrini, and Adil Ez-zetouni

7.1 INTRODUCTION

In a free-market global economy, the stock market is a key venue for people and businesses to exchange assets. The accessibility and possibility of interesting returns have made it a popular choice for investors and enterprises looking for funds (1). The volatile tendency of stock prices has long been considered a crucial concern within the economic domain (2). Stock prices are affected by a number of both internal and external variables, encompassing the national and foreign economic situations, international conditions, business prospects, financial statistics of publicly traded corporations, as well as the operations of the stock market (3, 4). Forecasting the stock market is among the most difficult functions in finance since stock markets are intrinsically unstable, chaotic, complex, nonlinear, dynamic, non-parametric, and non-stationary, rendering any prediction model susceptible to huge mistakes (5, 6). Furthermore, fluctuations in price are driven not just by past stock exchange data, but additionally by nonlinear variables, like political circumstances, shareholder attitude, and unforeseen incidents (7–10).

To address such challenges, many investigations have been done during the last decades to forecast different kinds of stock market time-series data. In fact, Linear approaches, like the autoregressive and moving average (ARMA) and autoregressive integrated moving average (ARIMA) methods, have demonstrated strong predictive performance for forecasting financial patterns. Nonetheless, classical statistical approaches presume the linearity of financial and macroeconomic time series (11), which is not applicable in real-life settings. Conversely, several machine learning approaches excel in capturing nonlinear associations inherent in the data (12, 13). Leveraging such techniques could be immensely advantageous in informing decision-making processes related to stock market investments (14). Similarly, deep-learning-based approaches have become prevalent in various financial applications including stock index and price, forecasting, risk management, portfolio optimization, trading techniques, and financial data processing. Particularly, Gated Recurrent Units (GRUs), Recurrent Neural Networks (RNNs), and

DOI: 10.1201/9781003478676-7

LSTMs have been developed specifically for handling time-series data and have been demonstrated to outperform classical time-series approaches, especially in situations where past events are crucial for forecasting future outcomes. Thereby, these advanced techniques have been widely employed in various tasks, including forecasting stock market index and translating languages (15, 16).

This research intends to construct an effective deep-learning algorithm that can forecast the Morocco All Share Index (MASI) movement within the financial market. This prediction will facilitate the determination of stock price trends, enabling the development of investment strategies and the identification of optimal trading opportunities. For this reason, our study compares three deep learning approaches, namely, GRU, RNN, and Convolutional Neural Network with Long Short-Term Memory (CNN-LSTM). The choice of using these models in forecasting the MASI index is motivated by their abilities to capture long-term dependencies, handle spatial features, extract hierarchical representations, balance memory retention, learn nonlinear relationships, and benefit from ensemble learning approaches.

The rest of this chapter is structured as follows. Section 7.2 exhibits the data and methods employed in this analysis. Section 7.3 presents the results of this study. Section 7.4 deals with the discussion. Section 7.5 sums up the present work with a conclusion.

7.2 DATA AND METHODS

7.2.1 Dataset description

In the present study, we used the real-time data of the Moroccan all share index (MASI), sourced directly from the Casablanca Stock Exchange. Data collection occurred, from January 2018 to March 2023, during the operating hours of the exchange, spanning Monday through Friday from 9:00 am to 3:30 pm local time, excluding weekends and holidays. Each row in the dataset corresponds to a specific date, with columns representing a particular sector index. These indices encompass a diverse array of economic sectors, including agri-food (AGRO), banking (BANK), chemistry (CHIM), Beverages (BOISS), electricity (ELEC), Engineering and Industrial Equipment (INBEI), Environmental Social and Governance (ESGI), Building and Construction Materials (BMC), MASI Gross Profitability (MASIR), insurance (ASSUR), MASI Net Profitability (MASRN), oil and gas (PG), as well as Transportation Services (SAC) and geopolitical risk (GPRD) (Table 7.1).

In purpose to train the model parameters, we use 80% of the data (Table 7.2),

$$Z = \left\{ \left\{ Z_{(t)}^{(i)} \right\}_{,} i = 1, \ldots 13, t = 1, 2, \ldots, N \right\}_{,}$$

Table 7.1 Stock prices

(a) First part of the table

Date	Close	AGRO	BANK	BMC	CHIM	BOISS	ELEC	INBEI
2018-01-02	12420.15	29091.65	14553.34	19593.63	4803.81	17065.74	2035.98	269.48
2018-01-03	12509.58	29892.43	14568.51	19563.96	5257.18	17065.74	2044.69	267.38
2018-01-04	12463.08	29597.36	14509.92	19057.30	5262.66	17274.86	2027.04	270.17
2018-01-05	12537.75	30079.29	14622.08	19070.48	5243.10	17274.86	2100.34	270.17
2018-01-08	12479.42	30043.26	14570.12	18757.21	5243.10	17274.86	2111.73	266.17

(b) Second part of the table

Date	ESGI	MASIR	ASSUR	MASRN	PG	SAC	GPRD
2018-01-02	1001.24	31421.07	5039.46	28404.66	15193.52	2476.92	188.690567
2018-01-03	1004.52	31647.33	5130.28	28609.20	15387.24	2453.85	138.023651
2018-01-04	1001.72	31529.69	5078.06	28502.85	15193.52	2438.46	146.357925
2018-01-05	1009.70	31718.58	5098.82	28673.61	15448.95	2461.54	66.517418
2018-01-08	1004.73	31571.03	5116.77	28540.22	15325.30	2431.54	82.172218

Table 7.2 Statistics summary

(a) First part of the table

	Close	AGRO	BANK	BMC	CHIM	BOISS	ELEC	INBEI
Count	1310.0	1310.0	1310.0	1310.0	1310.0	1310.0	1310.0	1310.0
Mean	11675.721	31243.034	13009.110	17206.235	4977.376	15570.081	2156.172	118.166
Std	1044.194	3650.272	1304.168	2264.786	1321.035	1510.739	219.858	61.710
Min	8987.89	23750.32	9591.42	11979.47	2665.02	12472.77	1575.64	35.63
Max	13991.47	40967.13	15707.05	22578.39	7535.90	19983.19	2793.30	275.58

(b) Second part of the table

	ESGI	MASIR	ASSUR	MASRN	PG	SAC	GPRD
Count	1310.0	1310.0	1310.0	1310.0	1310.0	1310.0	1310.0
Mean	900.218	32499.195	4566.013	28956.685	16109.047	3226.784	113.300148
Std	85.893	3433.725	537.872	2941.516	2900.119	603.685	59.429416
Min	669.59	24525.62	3525.08	21921.32	11120.44	2132.31	9.491598
Max	1086.56	40731.58	5701.54	36055.17	23639.05	4476.92	540.827393

where Z^i represents each feature that contributes to explaining the output variable, close price, which is a single variable:

$$Y = Y_t, \quad t = 1,\dots,N$$

7.2.2 Preprocessing

7.2.2.1 Outliers detection

Detecting outliers in a data set is an important step in data processing. Data points known as outliers are those that substantially deviate from the rest of the data set and can have a detrimental effect on the statistical analysis. After identifying the outliers, one must decide how to deal with them, whether by deletion, transformation, or imputation. The choice depends on the specific context of the analysis.

7.2.2.2 *Missing values detection*

Detecting missing values is another important step in the data processing process. Data may contain missing values for various reasons, such as entry errors, faulty sensors, or simply information not available at the time of collection.

Detecting missing values is crucial as it significantly influences the quality of the analysis. After detecting these values, we choose how to process them, whether by imputation (replacement with an estimated value) or by deleting the rows or columns concerned.

7.2.2.3 *Data division and normalizing*

Dividing data into separate training (80%) and testing (20%) sets is a fundamental strategy for evaluating the generalization ability of models. This choice of ratio is widely adopted in the literature for several reasons. On the one hand, allocating a larger proportion of the data to training allows models to better capture trends and patterns, essential to avoiding overfitting. On the other hand, reserving a portion of the data for testing ensures reliable evaluation of model performance on independent data, while maintaining a relatively small test set to ensure a variety of cases and accurate performance estimates. Moreover, the dataset was normalized using the min-max scale, which reduces all the variables into a range between [0,1]:

$$\bar{Z} = \frac{Z_t - Z_t^{min}}{Z_t^{max} - Z_t^{min}} \tag{7.1}$$

where Z_t represents the actual data of our feature variables and the Z_t^{max} and Z_t^{min} are the highest and the lowest price of each sample.

7.2.3 CNN-LSTM

7.2.3.1 *LSTM*

LSTM models (17) are RNNs (18, 19) that are particularly adept at learning and preserving long-term dependencies. Long-term dependencies can be learned and retained, making them useful for applications like financial time-series forecasting (20, 21). LSTM models consist of three gates: forgetting, input, and output. The forgetting gate determines if it should maintain/delete current data, while the input gate controls the amount of fresh data that is stored in memory. The output gate decides if the cell's current value contributes to the output. The value of the cell adds to the output. Forget gate: To choose which data to remove from LSTM memory, this gate employs a sigmoid function.

This result depends on the actual values of x_t and h_{t-1}. The value f_t, which ranges from 0 to 1, is the output of this gate and indicates whether the learned value is completely deleted or maintained. Here's how this output is calculated:

$$f_t = \sigma\ (W_{fh}[h_{t-1}], W_{fx}\ [x_t], b_f) \tag{7.2}$$

in which:

- f_t: Represents the forget gate's value at time t. This gate decides whether information from time step t–1 should be forgotten or preserved for time step t. The value of f_t is a real number between 0 and 1.
- W_{fh}: Refers to the weight matrix. It weights the hidden activations (h_{t-1}) from the preceding time step for the forget gate f_t.
- h_{t-1}: Represents the hidden activation vector from the preceding time step, t–1.
- W_{fx}: It is the weight matrix for the forgetting gate f_t, which weights the current time's inputs.
- x_t: The vector of input at the present point (t).
- B_f: The forgetting gate's partiality.
- σ: Represents the sigmoid function of activation.

The weighted total of inputs and hidden activations is transformed into a value between 0 and 1, which denotes the likelihood that each piece of information will be retained ($f_t = 1$) or forgotten

$$(f_t = 0)$$

Input Gate: This gate decides if the new information will be added to the LSTM memory. It consists of two layers, a sigmoid and a tanh layer. The sigmoid layer decides which values should be changed, and the tanh layer generates a vector of new candidate values for the LSTM memory. The output of both of these layers is obtained as follows:

$$i_t = \sigma(W_{ih}[h_{t-1}], W_{ix}[x_t], b_i)\ (3) c_t^{\sim} = tanh(W_{c_h}\left[h_{t-1}\right], W_{c_x}\left[x_t\right], b_c) \tag{7.3}$$

where i_t denotes if the value needs to be adjusted or not, and $c^{\sim}_t$ represents a vector of new potential values that ought to be stored in the LSTM memory. The joining of both of these layers provides an update to the LSTM memory. The present value is discarded using the forget gate by multiplying the previous value (i.e., c_{t-1}) and adding the new potential value ($i_t * \tilde{c}_t$). The mathematical equation goes as follows:

$$c_t = f_t * c_{t-1} + i_t * \tilde{c}_t \tag{7.4}$$

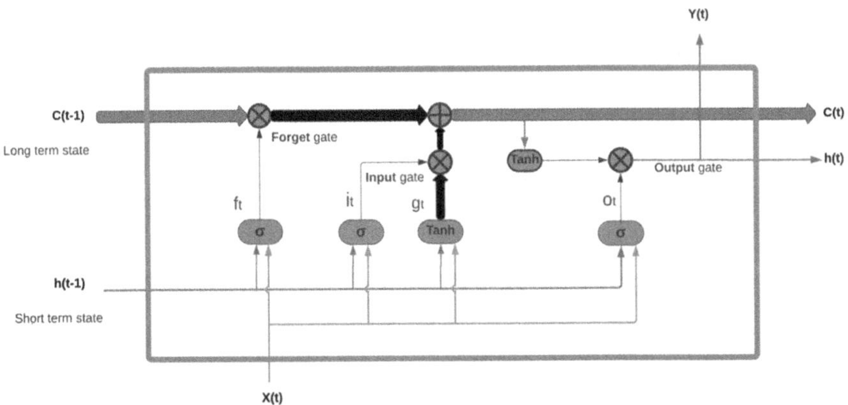

Figure 7.1 Diagram illustrating the structure of a long short-term memory (LSTM) network, highlighting the flow of data through various gates and cells (22).

With f_t constitutes the outcome of the forgetting gate, and is a number between 0 and 1, where 0 signifies entirely removing the value and 1 implies fully preserving it.

Output Gate: This gate starts with a sigmoid layer to identify which fraction of the LSTM memory adds to the result. Next, it uses a non-linear "tanh" function to transform values between –1 and 1. The outcome follows by multiplying by the result of a sigmoid layer. The equation below displays the equations used for calculating the final result:

$$o_t = \sigma(W_{oh}[h_{t-1}], W_{ox}[x_t], b_o)\ (6)h_t = o_t * tanh(c_t) \tag{7.5}$$

wherein o_t is the final value, and h_t is the representation as a number between –1 and 1.

This diagram, Figure 7.1, describes the architecture of an LSTM cell, which includes three distinct gates (the forget gate, the input gate, and the output gate) and the activation functions employed in each gate. The output formula can be found by conducting the operations depicted in the graph.

7.2.3.2 CNN

Lecun et al. proposed the CNN network concept in 1998 (23). CNN, a feed-forward neural network, performs well in image processing and Natural language processing (24). This approach is effective for forecasting time series. CNN's local perception and weight sharing enhance model learning efficiency by minimizing the number of parameters (25). CNN has a pair of layers: convolution and pooling. Formula (7.6) illustrates the computing formula for each convolution layer, which has several convolution kernels. Following the operation of the convolution layer, the data's features are extracted; however, because the extracted feature dimensions are rather large, this issue must be addressed in order to reduce training costs. To address this issue, a pooling layer is added following the convolution layer:

$$l_t = tanh(x_t * k_t + b_t) \tag{7.6}$$

where x_t is the input vector, k_t is the weight of the convolution kernel, *tanh* is the activation function, b_t denotes the kernel's bias, and l_t is the output value following convolution.

7.2.3.3 CNN-LSTM

CNN is widely used in feature engineering because of its capacity to concentrate on the most visible aspects in the field of vision. Because of its ability to expand in response to time sequence, LSTM is widely employed in time series analysis. A CNN-LSTM stock forecasting model (26) is designed based on their respective features. Figure 7.2 displays the model's main structure, which includes CNN and LSTM with input, one-dimensional convolution, pooling, an LSTM hidden layer, and full connectivity layers.

7.2.4 GRU network

The GRU network, introduced by Cho et al. (27), is a somewhat dramatic variation of LSTM which has one fewer gate. In the GRU network, there are just two gates: the reset gate and the update gate. The reset gate regulates how much the prior memory is disregarded and blends the incoming input with the old memory. More memory is disregarded the lower the reset gate's value. The amount of memory that is reserved beforehand is ascertained using the update gate. The update gate will produce more information the higher its value. When we set the update gate values to 0 and all of the reset gate values

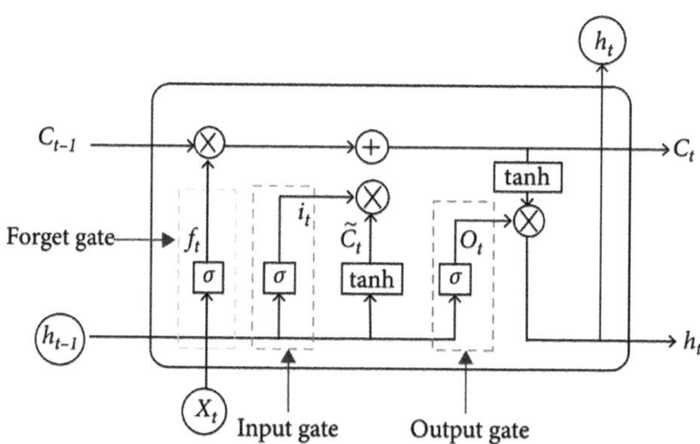

Figure 7.2 Structural diagram showing the integration of convolutional neural networks (CNN) with long short-term memory (LSTM) networks, detailing the layers and connections between them (26).

to 1, we obtain the general neural network model. The following equations are the mathematical explanations of the GRU cell functions:

$$z_t = \sigma(W_z \cdot [h_{t-1}, x_t]) \tag{7.7}$$

$$r_t = \sigma\left(W_r \cdot \left[h_{t-1}, x_t\right]\right) \tag{7.8}$$

$$\tilde{h}_t = \tanh\left(W_{\tilde{h}_t}\left[r_t^* h_{t-1}, x_t\right]\right) \tag{7.9}$$

$$h_t = \left(1 - z_t\right) * h_{t-1} + z_t * \tilde{h}_t \left(12\right) y_t = \sigma\left(W_o \cdot h_t\right) \tag{7.10}$$

The update gate (z_t) and forgetting gate (r_t) in the GRU are obtained from the input (x_t) and the hidden layer (h_{t-1}) at the previous time and are multiplied by the weight matrices W_z and W_r, respectively, same like the forgetting gate and input gate in the LSTM. Through the use of the sigmoid activation function $\sigma(x)$ on $h(t)$, the final output is compressed to $0 - 1$.

7.2.5 Recurrent neural networks

RNN belongs to the class of neural networks (28) in which there is a recurring connection between the units. This enables them to process the series of inputs using their internal memory. This enables applications such as speech recognition, text production, stock market, and handwritten recognition. This study makes use of neural networks since the stock data requires long-term dependencies (29) to be taken into account.

7.2.6 SHAP values method

Machine learning models have become increasingly complex, making it challenging to understand the reasoning behind their predictions. SHAP (SHapley Additive exPlanations) values offer a solution by providing a method to explain the contribution of each feature to the model's predictions. Indeed, this method is applied in this research in order to understand the influence of individual features on the model's predictions (30).

7.2.7 Evaluation metrics

In this research study, we appraised the models using the Mean Absolute Percentage Error (31, 32) (MAPE), the r-square (33) (r^2), and the root mean squared error (RMSE) (34).

- The MAPE is calculated as follows:

$$\text{MAPE} = \frac{1}{N}\sum_{i=1}^{N}\left|\frac{Z_{a,i} - Z_{p,i}}{Z_{a,i}}\right| * 100 \tag{7.11}$$

where N is the number of observations, $z_{a,i}$ are the actual values, and $z_{p,i}$ are the predicted values, the closer this percentage is to 0, the less the model make errors and the better it is.

- The r-square is computed as follows:

$$r^2 = 1 - \frac{\frac{\sum_{i=1}^{N}(Z_{a,i} - Z_{p,i})^2}{N}}{\frac{\sum_{i=1}^{N}(\bar{Z} - Z_{p,i})^2}{N}} \tag{7.12}$$

where N is the number of observations, $z_{a,i}$ are the actual values, $z_{p,i}$ are the predicted values and z is the average value, the r-square is between 0 and 1 the closer it is to 1 the better our model is.

- The formula of the root mean square error is as follows:

$$\text{RMSE} = \sqrt{\frac{1}{N}\sum_{i=1}^{N}(Z_{a,i} - Z_{p,i})^2} \tag{7.13}$$

where N is the number of observations, $z_{a,i}$ are the actual values, $z_{p,i}$ are the predicted values.

7.3 RESULTS

7.3.1 CNN-LSTM

The CNN-LSTM model's forecasting results for the MASI index are quite encouraging, demonstrating the model's capacity to anticipate future values. The test set's Root Mean Square Error (RMSE) is 127.27, meaning that the model's predictions often differ by this amount from the actual values. In this instance, The RMSE indicates that the model's forecasts are pretty close to the actual values. A lower RMSE denotes improved accuracy.

The coefficient of determination, or r^2, is a crucial regression analysis metric that expresses the percentage of the dependent variable's variance (in this case, the MASI index) that can be predicted from the independent variables (input data). With a relatively high r^2 value of 0.9796, the CNN-LSTM model can account for almost 97.96% of the variation in the MASI index using the input data. This suggests that there is a good correlation between the target variable and the input features, which results in precise predictions.

In addition, the model's prediction accuracy is indicated as a percentage via the MAPE, which is 0.8857. A MAPE less than 1 signifies that the model's predictions are inaccurate by less than 1% on average, which is an excellent outcome. With forecasts that are consistently close to the actual values, this implies that the CNN-LSTM model is a credible forecaster for the MASI index. All things considered, these findings show that the CNN-LSTM model is a reliable and accurate method for MASI index forecasting,

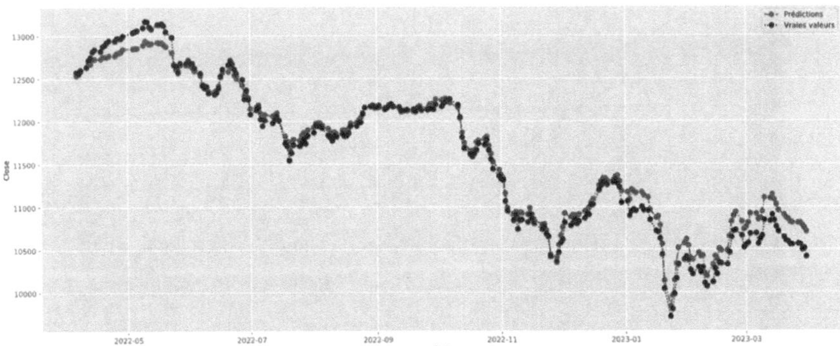

Figure 7.3 Graph depicting the closing price predictions made using a CNN-LSTM model, comparing the predicted values with the actual values over time.

offering insightful information for investment strategies and decision-making (Figure 7.3).

7.3.2 GRU

The performance metrics collected demonstrate the high degree of accuracy and dependability of the MASI index predicting outcomes utilizing the GRU model.

First off, the test set's Root Mean Square Error (RMSE) is 155.03. The average magnitude of the discrepancies between the actual MASI index values and the projected values from the GRU model is represented by this metric. In this instance, the RMSE of 155.03 indicates that the GRU model's predictions are reasonably near to the true values. A lower RMSE denotes improved accuracy. Furthermore, based on the input data, the coefficient of determination, or r^2, expresses how well the GRU model explains the variation in the MASI index. The model's r^2 value of 0.9697 indicates a significant correlation between the input features and the target variable, accounting for about 96.97% of the variability in the MASI index. The GRU model's ability to capture the underlying patterns and trends in the MASI index data is indicated by the high r^2 value.

Finally, 1.1345 is the MAPE for the GRU model. As a proportion of the actual values, the MAPE calculates how accurate the model's predictions are. A MAPE of roughly 1.1345 implies that, on average, the GRU model's predictions depart from the actual values by roughly 1.1345%. A lower MAPE indicates better accuracy. This degree of precision is fairly high, demonstrating the GRU model's dependability in MASI index prediction. Therefore, with a high r^2 value, a manageable MAPE, and a comparatively low RMSE, the GRU model shows good predicting performance. These findings show that the GRU model is a useful tool for financial forecasting and decision-making since it is good at identifying and predicting the trends and patterns in the MASI index data (Figure 7.4).

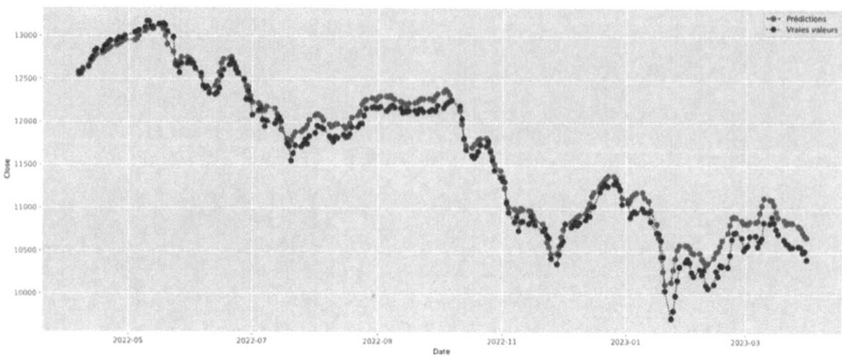

Figure 7.4 Graph showing the closing price predictions using a gated recurrent unit (GRU) model, with a comparison of the predicted prices against the true values.

7.3.3 RNN

Excellent performance is also demonstrated by the MASI index forecasting outcomes using a typical RNN model. The average size of the discrepancies between the actual MASI index values and the anticipated values from the RNN model is shown by the RMSE of 108.77. In this instance, the RMSE value of 108.77 indicates that the RNN model's predictions are reasonably near to the true values. A lower RMSE denotes better accuracy. The accuracy of the model's predictions as a percentage of the actual values is indicated by the MAPE of 0.7817. A MAPE of roughly 0.7817 suggests that, on average, the RNN model's predictions differ from the actual values by roughly 0.7817%. A MAPE of close to 1 indicates good accuracy. This excellent degree of accuracy shows how trustworthy the RNN model is in predicting the MASI index. In addition, the coefficient of determination (r^2) is a crucial statistic that expresses how much of the variance in the MASI index can be accounted for by the input features that the RNN model uses. Based on the input data, the model can account for about 98.51% of the variability in the MASI index, with a r^2 value of 0.9851. The significant correlation between the input characteristics and the target variable, as suggested by the high r^2 value, demonstrates that the RNN model efficiently captures the underlying patterns and trends in the MASI index data. Obviously, as observed, the RNN model has excellent predicting performance, as evidenced by its low RMSE, high r^2 value, and excellent MAPE. These findings demonstrate the MASI index prediction accuracy and dependability of the RNN model, which makes it a useful tool for financial forecasting and decision-making activities (Figure 7.5).

7.3.3.1 Interpreting RNN model predictions using SHAP values

The features that contribute most and least to our best model's prediction are illustrated in Figure 7.6, considering both the impact (SHAP value) and the feature's value itself.

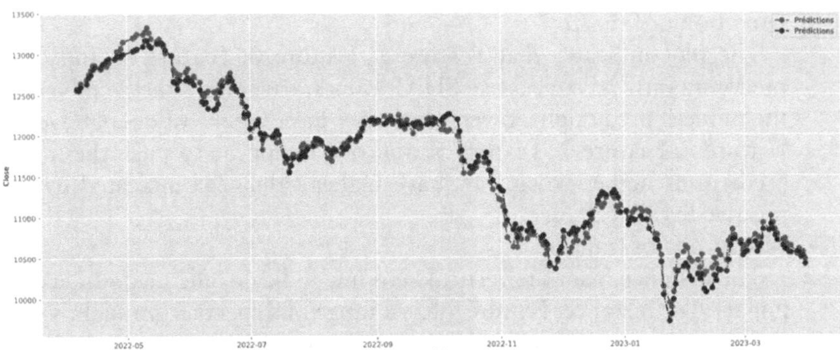

Figure 7.5 Graph illustrating the closing price predictions using a recurrent neural network (RNN), displaying the relationship between predicted and actual closing prices.

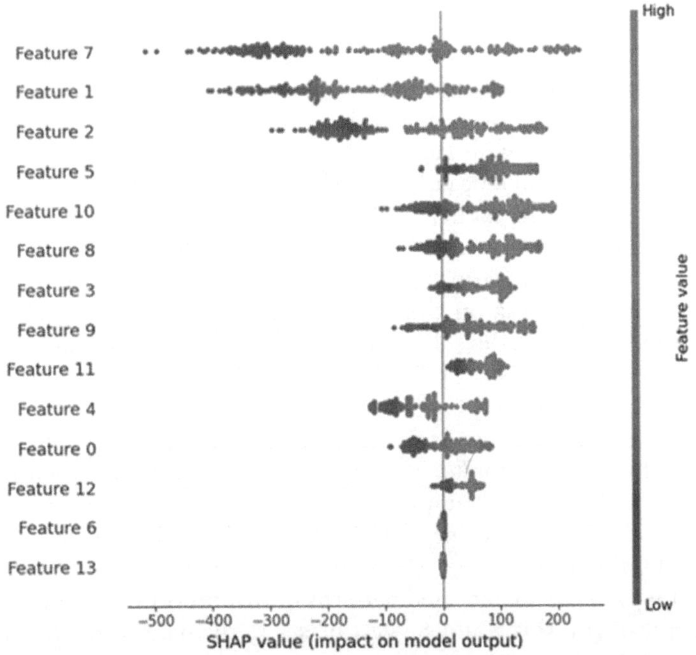

Figure 7.6 Bar chart representing feature importance calculated using SHAP values for a recurrent neural network (RNN) model, showing the impact of different features on the model's predictions.

- Feature Importance:

  The order of features on the y-axis indicates their importance in terms of impact on the model output. Feature 7 is at the top, suggesting it has the highest overall impact, whether positive or negative, across all predictions.

- Direction of Impact:
  The plot indicates that Feature 1, Feature 0, Feature 6, and others predominantly have negative SHAP values, suggesting they tend to push the model's predictions lower when they have higher values. Conversely, Feature 7, Feature 2, Feature 5, and others appear to push the model's predictions higher when they have higher values, as indicated by their positive SHAP values.
- Magnitude of Impact:
  The distance from the vertical zero line indicates the magnitude of the impact. For instance, Feature 7 has a range of impacts from slightly negative to quite positive, with a concentration of positive impacts suggesting it often contributes to increasing the prediction value. Features with a tight cluster of points close to the zero line, like Feature 4 and Feature 13, seem to have a smaller impact on the model output as their SHAP values are close to 0.
- Feature Value Correlation:
  The color intensity shows a high feature value (pink) or a low feature value (blue). For example, high values of Feature 7 contribute to a higher prediction value, while low values of Feature 1 contribute to a lower prediction value. This suggests a positive correlation with the outcome for Feature 7 and a negative correlation for Feature 1.
- Consistency of Impact:
  Features such as Feature 10 and Feature 8 have a wide spread of SHAP values, indicating that their impact on the model output is highly variable depending on the specific context or data point. Conversely, Feature 4 and Feature 13 have a very narrow spread, indicating they have a more consistent but modest impact on the model's output.
- Conflict in Impact:
  Some features show a mix of both positive and negative SHAP values for different data points, indicating that the direction of their impact on the model's output is not consistent. This could be due to interactions with other features or non-linear effects within the model.
- Potential Bias:
  If we see any unexpected patterns or disproportionate impacts of certain features, this could suggest biases in the data or model that might require further investigation.

7.3.3.2 Forecasting future stock prices of MASI in real-time

Figure 7.7 depicts the forecasting result of the MASI index from 13-03-2023 to 20-122023. Then we can notice the existence of an ascending trend in the forecasting MASI index starting from 10600 DH in 14-03-2023 to approximately 11800 DH in 20-03-2023 reaching its peak in late October 2023. While comparing these results with the actual values in the Casablanca stock exchange, we deduce that our model fits perfectly the expectations and forecasts the values of the MASI index flawlessly.

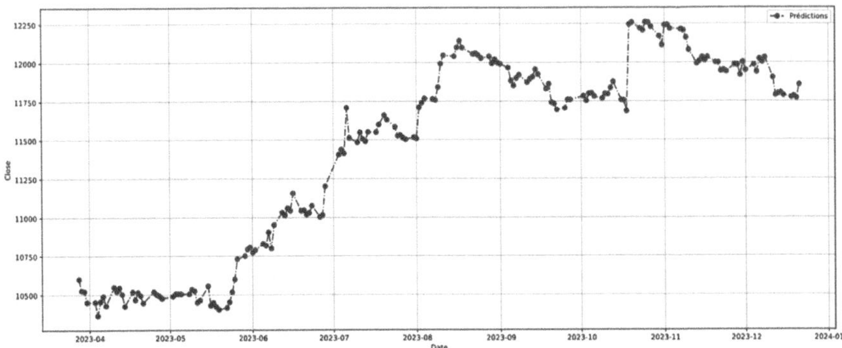

Figure 7.7 Graph forecasting the MASI index values from March 13, 2023, to December 20, 2023, indicating the predicted trend over this period.

7.4 DISCUSSION

The present study's outcomes indicate that the implementation of RNN demonstrates superior performance concerning the MASI index, exhibiting a noteworthy R-squared value of 98.51%, a MAPE of 0.7817, and a Root Mean Squared Error (RMSE) of 108.77.

Upon juxtaposing our findings with those delineated in prior works such as (35, 36), and (37), our analysis corroborates that our premier model, RNN, garners superior results in terms of R-squared and MAPE in contrast to alternative models. Moreover, with regard to RMSE, our RNN model surpasses the optimal algorithm identified by (35), yet it is outstripped by the outcomes reported by (36) and (37) (refer to Table 7.3).

- The models used in Ireland 2019, including MA, SVR, and LSTM, exhibit varying levels of predictive accuracy. The LSTM model stands out with an R^2 value of 83.00%, indicating that it explains 83.00% of the variance in the data. Additionally, the low RMSE and MAPE values suggest relatively accurate predictions.

Table 7.3 ML and DL model comparison analysis in historical financial time series forecasts

Markets	*Year*	*Comparative models*	*RMSE*	*MAPE (%)*	R^2*(%)*	*Reference*
Ireland	2019	MA, SVR, LSTM,	347.46	1.03	83.00	(35)
China	2020	MLP, CNN [3], RNN LSTM, CNN-RNN	39.688	—	96.50	(36)
American	2022	ARIMA, XGBoost, LSTM	6.1010	3.800	96.10	(37)
Morocco	2023	CNN-LSTM, GRU, RNN	108.77	0.7817	98.51	Our Study

- In China 2020, a combination of MLP, CNN, RNN, LSTM, and CNN-RNN models was employed. The R^2 value of 96.50% indicates a high level of predictive power, although the RMSE value suggests moderate error in predictions.
- The models utilized in American 2022, such as ARIMA, XGBoost, and LSTM, demonstrate strong predictive performance with an R^2 value of 96.10%. The low RMSE and MAPE values further support the accuracy of predictions across these models.
- Morocco 2023 employed CNN-LSTM, GRU, and RNN models, achieving an impressive R^2 value of 98.51%. The low RMSE and MAPE values indicate accurate predictions and excellent explanatory power.

7.5 CONCLUSION

This present chapter sought to assess the effectiveness of several deep learning algorithms in forecasting the Morocco All Share Index (MASI) inside the financial market. We implemented three deep learning models: CNN-LSTM, GRU, and RNN, and rigorously tested their predictive performance.

Through our analysis, we observed that the RNN model outperformed the other algorithms, demonstrating a substantial improvement in forecasting accuracy for the MASI index. This finding underscores the potential of RNN-based approaches in capturing the complex temporal dependencies and patterns inherent in financial time series data, thereby enhancing predictive capabilities. Despite the success of RNN in our study, it is important to note that the field of deep learning for financial forecasting is continuously evolving. Future research endeavors could delve deeper into optimizing hyperparameters, exploring ensemble methods that combine different deep learning architectures, and integrating additional data sources to further refine forecasting accuracy. Furthermore, addressing limitations such as data quality, market volatility, and model interpretability remains crucial for advancing the reliability and applicability of deep learning techniques in financial forecasting.

Overall, our study contributes to the growing body of research on leveraging deep learning algorithms for stock market prediction and underscores the potential of RNN models in accurately anticipating movements in the Morocco All Shares Index, thereby offering valuable insights for investors, analysts, and financial decision-makers.

REFERENCES

[1] K. K. Yun, S. W. Yoon, and D. Won. *Prediction of stock price direction using a hybrid GA-XGBoost algorithm with a three-stage feature engineering process. Expert Systems with Applications*, vol. 186, p. 115716, 2021.
[2] R. Vanaga, and B. Sloka. *Financial and capital market commission financing: Aspects and challenges. Journal of Logistics, Informatics and Service Science*, vol. 7, no. 1, pp. 17–30, 2020.

[3] L. L. Zhang, and H. Kim. *The influence of financial service characteristics on use intention through customer satisfaction with mobile fintech. Journal of System and Management Sciences*, vol. 10, no. 2, pp. 82–94, 2020.
[4] L. Badea, V. Ionescu, and A.-A. Guzun. *What is the causal relationship between Stoxx Europe 600 sectors? But between large firms and small firms? Economic Computation & Economic Cybernetics Studies & Research*, vol. 53, no. 3, 2019.
[5] T. Z. Tan, C. Quek, and G. Ng. *Brain-inspired genetic complementary learning for stock market prediction. 2005 IEEE Congress on Evolutionary Computation*, vol. 3, pp. 2653–2660, 2005.
[6] J.-Z. Wang, J.-J. Wang, Z.-G. Zhang, and S.-P. Guo. *Forecasting stock indices with back propagation neural network. Expert Systems with Applications*, vol. 38, no. 11, pp. 14346–14355, 2011.
[7] E. F. Fama. *The behavior of stock-market prices. The Journal of Business*, vol. 38, no. 1, pp. 34–105, 1965.
[8] X. Zhang, X. Liang, A. Zhiyuli, S. Zhang, R. Xu, and B. Wu. *At-LSTM: An attention-based LSTM model for financial time series prediction. IOP Conference Series: Materials Science and Engineering*, vol. 569, no. 5, p. 052037, 2019.
[9] R. Shields, S. Ajour El Zein, and N. Vila Brunet. *An analysis on the NASDAQ's potential for sustainable investment practices during the financial shock from Covid-19. Sustainability*, vol. 13, no. 7, p. 3748, 2021.
[10] M. K. Daradkeh. *A hybrid data analytics framework with sentiment convergence and multi-feature fusion for stock trend prediction. Electronics*, vol. 11, no. 2, p. 250, 2022.
[11] I. Bouayad, J. Zahir, and A. Ez zetouni. *Nowcasting and forecasting Morocco GDP growth using Google trends data. IFAC-PapersOnLine*, vol. 55, pp. 3280–3285, October 2022. doi: 10.1016/j.ifacol.2022.10.129
[12] S. Abrishami, M. Turek, A. R. Choudhury, and P. Kumar. *Enhancing profit by predicting stock prices using deep neural networks.* In *2019 IEEE 31st International Conference on Tools with Artificial Intelligence (ICTAI)*, pp. 1551–1556, 2019.
[13] H. Oukhouya, and K. E. Himdi. *A comparative study of ARIMA, SVMs, and LSTM models in forecasting the Moroccan stock market. International Journal of Simulation and Process Modelling*, vol. 20, no. 2, pp. 125–143, 2023.
[14] S. Aggarwal, and S. Aggarwal. *Deep investment in financial markets using deep learning models. International Journal of Computer Applications*, vol. 162, no. 2, pp. 40–43, 2017.
[15] B. Ren. *The use of machine translation algorithm based on residual and LSTM neural network in translation teaching. PLoS One*, vol. 15, no. 11, p. e0240663, 2020.
[16] H. N. Bhandari, B. Rimal, N. Pokhrel, R. Rimal, K. R. Dahal, and R. K. C. Khatri. *Predicting stock market index using LSTM. Machine Learning with Applications*, vol. 9, p. 100320, 2022.
[17] S. Hochreiter, and J. Schmidhuber. *Long short-term memory. MIT Press*, vol. 9, no. 8, pp. 1735–1780, 1997.
[18] V.-D. Ta, C.-M. Liu, and D. A. Tadesse. *Portfolio optimization-based stock prediction using long-short term memory network in quantitative trading. Applied Sciences*, vol. 10, no. 2, pp. 437–457, 2020.
[19] O. Zarrad, M. A. Hajjaji, and M. N. Mansouri. *Hardware implementation of hybrid wind-solar energy system for pumping water based on artificial neural network controller. Studies in Informatics and Control*, vol. 28, no. 1, pp. 35–44, 2019.

[20] S. Siami-Namini, N. Tavakoli, and A. Siami Namin. *A comparison of ARIMA and LSTM in forecasting time series*. In *2018 17th IEEE International Conference on Machine Learning and Applications (ICMLA)*, pp. 1394–1401, IEEE, 2018.
[21] K. Pawar, R. S. Jalem, and V. Tiwari. *Stock market price prediction using LSTM RNN*. In V. Rathore, M. Worring, D. Mishra, A. Joshi, S. Maheshwari (eds), *Emerging Trends in Expert Applications and Security*, Advances in Intelligent Systems and Computing, vol. 841. Springer, Singapore, 2019.
[22] C. Olah. *Understanding LSTM networks* [J/OL]. colah's blog. 2015-08-27. Retrieved 2020-05-18 from https://colah.github.io/posts/2015-08-Understanding-LSTMs, 2015.
[23] Y. Lecun, L. Bottou, Y. Bengio, and P. Haffner. *Gradient-based learning applied to document recognition. Proceedings of the IEEE*, vol. 86, no. 11, pp. 2278–2324, 1998.
[24] B. S. Kim, and T. G. Kim. *Cooperation of simulation and data model for performance analysis of complex systems. International Journal of Simulation Modelling*, vol. 18, no. 4, pp. 608–619, 2019.
[25] L. Qin, N. Yu, and D. Zhao. *Applying the convolutional neural network deep learning technology to behavioral recognition in intelligent video. Tehnicki VjesnikTechnical Gazette*, vol. 25, no. 2, pp. 528–535, 2018.
[26] W. Lu, J. Li, Y. Li, A. Sun, and J. Wang. *A CNN-LSTM-based model to forecast stock prices. Complexity*, vol. 2020, Article ID 6622927, 10 pages, 2020.
[27] K. Cho, V. Merrienboer, and C. Gulcehre. *Learning phrase representations using RNN encoder-decoder for statistical machine translation*. arXiv:1502.08029, February 2015.
[28] A. Dewan, and M. Sharma. *Prediction of heart disease using a hybrid technique in data mining classification*. In *Computing for Sustainable Global Development (INDIACom), 2015 2nd International Conference on*, pp. 704–706, IEEE, 2015.
[29] H. Sak, A. Senior, and F. Beaufays. *Long short-term memory recurrent neural network architectures for large scale acoustic modeling*. In *Fifteenth Annual Conference of the International Speech Communication Association*, 2014.
[30] Y. Nohara, K. Matsumoto, H. Soejima, and N. Nakashima. *Explanation of machine learning models using Shapley additive explanation and application for real data in hospital. Computer Methods and Programs in Biomedicine*, vol. 214, p. 106584, 2022.
[31] J. McKenzie. *Mean absolute percentage error and bias in economic forecasting. Economics Letters*, vol. 113, no. 3, pp. 259–262, 2011.
[32] A. de Myttenaere, B. Golden, B. Le Grand, and F. Rossi. *Mean absolute percentage error for regression models. Neurocomputing*, vol. 192, pp. 38–48, 2016.
[33] D. Chicco, M. J. Warrens, and G. Jurman. *The coefficient of determination R-squared is more informative than SMAPE, MAE, MAPE, MSE and RMSE in regression analysis evaluation. PeerJ Computer Science*, vol. 7, p. e623, 2021.
[34] M. V. Shcherbakov, A. Brebels, N. L. Shcherbakova, A. P. Tyukov, T. A. Janovsky, and A. E. Kamaev. *A survey of forecast error measures. World Applied Sciences Journal*, vol. 24, no. 24, pp. 171–176, 2013.
[35] S. K. Lakshminarayanan, and J. P. McCrae. *A comparative study of SVM and LSTM deep learning algorithms for stock market prediction*. In *Proceedings of the AICS*, Wuhan, China, *12–13 July 2019*, pp. 446–457, 2019.
[36] W. Lu, J. Li, Y. Li, A. Sun, and J. Wang. *A CNN-LSTM-based model to forecast stock prices. Complexity*, vol. 2020, p. 6622927, 2020.
[37] G. Goverdhan, S. Khare, and R. Manoov. *Time series prediction: Comparative study of ML models in the stock market. Research Square*, Version, 2022.

Chapter 8

A comprehensive review of artificial intelligence techniques for timely and accurate prediction of Down syndrome

Loubna Taidi and Sokaina El Khamlichi

8.1 INTRODUCTION

Pregnancy and childbirth can be the most exciting times in a woman's life [1]. However, they can also be the most stressful, as many parents have questions about their unborn baby's health [2]. It is recommended that all expectant mothers screen in the first trimester to determine whether their unborn child has any of the screened conditions [3]. Down syndrome (DS) is a topic that comes up early in every pregnancy [4].

DS, also known as trisomy 21, is one of the most common congenital developmental disabilities caused by chromosomal abnormalities in humans [5]. Autosomal duplication, which results in an extra copy of chromosome 2 or chromosome 21, is the main cause of DS [6]. A wide range of clinical characteristics, such as intellectual disability, small stature, flat face, flat nasal bridge, pronounced epicanthic folds, up-slanting palpebral fissures, and protruding tongue, are caused by that additional chromosome, which impacts most body systems [7].

Early detection is key to preventing future health issues and providing patients with lifetime benefits such as speech, physical, cardiac, and neurological therapy [8]. Prenatal DS screening is an essential part of antenatal treatment and should be made available to all women, regardless of their age or circumstances [9]. While screening tests can only determine your likelihood of having a child with DS, they can assist you in choosing more specialized diagnostic tests [10].

Recently, there is potential for growth in several domains, thanks to recent developments in deep learning (DL) and computer vision. The results of tasks like item identification, localization, recognition, and segmentation based on public datasets have significantly improved [11]. In 2012, a prenatal test reached the market, quickly transforming the prenatal testing paradigm. Using cell-free fetal DNA that is expressed in the mother's blood between weeks 11 and 22 of pregnancy and comes from the placenta, novel non-invasive prenatal testing (NIPT) may be a promising development [12]. Since NIPT is predicated on the examination of cell-free fetal DNA for chromosomal disorders, it provides a stage in between serum screening and invasive

DOI: 10.1201/9781003478676-8

diagnostic testing [13, 14]. Notwithstanding the aforementioned possible advantages of NIPT, there are some important drawbacks. Test results that are falsely positive or negative do happen, and in clinical practice, they can happen more frequently than in tightly monitored, small-scale clinical studies. Additionally, because the test examines the mother's and the child's DNA, it can identify a genetic disorder the mother was unaware she had [15, 16].

Furthermore, there is significant interest in applying sophisticated computational methods, such as machine learning (ML), DL, and transfer learning, for the early diagnosis of DS [17]. In fact, it allows the timely intervention and support for individuals with this condition as well as their families. Additionally, the early identification of DS in pregnancy can help expectant parents in making informed decisions about their pregnancy and access necessary resources and support services. Furthermore, it enables healthcare providers to initiate appropriate medical management and developmental interventions, ultimately improving outcomes for individuals with DS. Besides, these artificial intelligence (AI)-powered screening protocols have the potential to reduce healthcare costs associated with DS screening by streamlining processes, reducing the need for unnecessary tests, and optimizing resource allocation. This cost-effectiveness makes screening more accessible to a broader population, ensuring equitable access to prenatal care.

In this context, the present chapter contributes significantly to the field of DS diagnosis. This comprehensive review intends to explore the state-of-the-art of the application of AI for the early detection and diagnosis of Down syndrome. Our analysis strives to lessen the financial and societal costs related to prenatal diagnosis, through synthesizing existing knowledge and advancements in AI techniques. Thereby, it intends to boost this field toward effective, accessible, and affordable diagnostic solutions for DS.

To guarantee a comprehensive scope of the subject, we utilized a systematic search methodology to identify relevant studies relating to the prediction of DS using AI, ML, and NIPT methods. The search was conducted across multiple scientific databases, including PubMed, IEEE Xplore, and Google Scholar. A systematic approach was adopted, employing specific keywords and Boolean operators such as "Down syndrome prediction," AI, ML, and "NIPT method." These keywords were selected based on their relevance to the research topic and to ensure the retrieval of all pertinent literature. The search strategy was designed to encompass a broad range of studies published in peer-reviewed journals and conference proceedings. Furthermore, manual searches of reference lists from relevant articles were conducted to identify additional studies that may have been missed during the initial database search. The inclusion criteria for studies considered in this review were predetermined, focusing on research that employed AI, ML, or NIPT methods for the prediction of DS in prenatal or clinical settings. Duplicate studies were removed, and the remaining articles underwent a rigorous screening process based on their title, abstract, and full text. Finally, data extraction and synthesis were performed to analyze the findings of the included studies and draw meaningful conclusions regarding the current state of research in this field.

The rest of this chapter is organized as follows. Section 8.2 illustrates the importance of AI in DS detection. Section 8.3 presents the existing research which is divided into ML approaches for DS diagnosis and DL methods used for DS diagnosis. Section 8.4 deals with the discussion. Section 8.5 sums up this work with a conclusion.

8.2 IMPORTANCE OF ARTIFICIAL INTELLIGENCE IN DOWN SYNDROME DETECTION

Over the past 50 years, AI has been developed sustainably since it is widely used in many fields and has a robust applicability [18]. AI is the field of computers that will do tasks that humans can complete, like classifying and identifying images, translating between languages, and identifying and solving specific problems. AI has been used in many fields to produce more meaningful and better data analysis [19, 20].

ML is the most widely used subfield of AI [21]. In computer science, ML is the study of algorithms and approaches for automating the solving of complicated issues that are challenging to solve using traditional programming techniques [22]. Medical science is increasingly using DL and ML, especially in the domains of visual, aural, and linguistic data [23]. Large, complicated datasets can be analyzed by ML algorithms, which can then spot patterns and trends that humans find difficult to identify as illnesses [24]. The capacity to properly anticipate diseases can assist medical personnel to take proactive actions to prevent and treat diseases early, resulting in better health outcomes for patients [25]. In this context we can understand that ML techniques can be used to solve problems with high-dimensional datasets. In several fields, ML makes a greater contribution. Situated on the cusp of a significant change in healthcare epidemiology, many of the intricate models leverage the greater training data that is now available [26].

Diagnosis of chromosomal abnormalities in fetus is one of the most important challenges in modern perinatology [27]. Recently, advanced clinical testing and purification methods have ameliorated the performance of NIPT and enabled the use of new techniques [28]. With advances in AI, statistical ML can guarantee accuracy and exceed manual effectiveness in analyzing and evaluating NIPT results. The use of this technique in clinical settings presents NIPT with significant promise for the screening of aneuploidy [29]. Hence, using this new technique, corrective action can be taken, and irreversible outcomes can be avoided with the early and accurate discovery of genetic diseases. Thus, it is crucial to have precise and low-risk diagnostic techniques to identify these illnesses at the fetal stage.

Moreover, timely diagnosis and ongoing assistance are essential for people with DS to attain their full potential while leading satisfying lives [30].

Primary intervention programs use a multidisciplinary strategy to give complete assistance in problems like cognition, motor abilities, communication, and social-psychological advancement [30, 31]. Personalized education programs customize educational objectives and adaptations based on every child's specific requirements, encouraging comprehensive education and development of abilities [30]. Healthcare management, involving periodic screenings and preventive care for related health issues, promotes the best wellness results [32–34].

8.3 EXISTING RESEARCH

8.3.1 Machine learning approaches for DS diagnosis

According to Yang et al. [35], using an ML algorithm in conjunction with several Z tests might result in a more accurate prediction using NIPT data. Yang et al. obtained features from well-known samples and scaled them for support vector machine (SVM) discriminating model using quality control and clinical sign indexes in conjunction with numerous Z values. When the team used the trained model to forecast the unidentified samples, it has significantly improved. This technique achieved 100% accuracy on all three chromosomes in 4752 certified NIPT data, including 151 data that were classified as unclassified by a method based on one Z value. Furthermore, they used an ML model to accurate erroneous positives and false negatives [35].

Additionally, Koivu et al. [36] examined ML methods to enhance the effectiveness of first-trimester DS screening. They experimented multiple classification techniques using two real-world datasets, and then the team compared the performance of the implemented models to a predicate technique, a program for assessing commercial risk, after they were evaluated using a third group of actual data. With the test data, the area under the curve for the top-performing deep neural network model was 0.96, with a 78% detection rate and a 1% false positive rate. The SVM model generated an area under the curve of 0.95, a detection rate of 61%, and a false positive rate of 1% using the identical test data. As a result, Koivu et al. found that the most effective model for a SVM performed somewhat worse than the predicate technique, while an optimized deep neural network model produced better identification rates with identical false positive rate or the same identification rates but a much lower probability of false positive rate [36].

In other way, Neocleous et al. [37] attempted to use AI classification techniques to evaluate the danger for fetal trisomy 21 (T21) and more chromosomal malformation during 11 and 13 weeks of pregnancy. First, they use 72054 euploid gestations, 295 T21 instances, and 305 samples of other chromosomal abnormalities (OCA) to train the artificial neural networks (ANNs). After that, they divide the situations into two groups: "risk" and "no-risk." The "no-risk" cases are closed, and the "risk" cases are sent to

Stage 2 for additional review, where they categorize them into three risk categories: "no-risk," "moderate-risk," and "high-risk." A total of 36,328 pregnancies that were not known to the system led to the classification of 17,512 euploid, 2 T21, and 18 other chromosomal anomalies as "no-risk" in the first stage. Moreover, the team reallocate the remaining 18,796 instances (51.4% FPR) in Stage 2, where 10,464 euploid, 187 euploid, 2 T21 and 2 OCA, 50 T21 and 52 OCA, 83 T21 and 61 OCA, and 7895 euploid are categorized as "no-risk," "moderate-risk," and "high-risk." In the next stage, T21's sensitivity and specificity are 97.1% and 99.5%, respectively, supposing that a cell-free DNA test can distinguish between every euploid and aneuploid instance [37].

Else, Catic et al. [38] demonstrated that maternal serum screening results in the first trimester, ultrasonographic results, and patient demographics can all be utilized to effectively classify five aneuploidy syndromes using a professional diagnostic framework built on ANNs. They included in the database information gathered from 2,500 expectant mothers who visited the "Mehmedbasic" Institute of Gynecology, Infertility, and Perinatology for standard prenatal treatment. In the initial trimester, the team applied screening tests to all women to determine levels of free beta-human chorionic gonadotropin (β-hCG) and maternal blood pregnancy-associated plasma protein-A (PAPP-A). Additionally, they used ultrasonography to detect the thickness of the fetal nuchal translucency and whether the nasal bone is present or not. They looked at the ANNs using linear feedforward and feedback for different distributions of training data and hidden layer neuron counts. In terms of prediction power for each of the five classifications of aneuploidy prenatal syndrome, Catic et al. found that feedback neural network design outperformed feedforward neural network architecture. A feedforward neural network with 15 hidden layer neurons attained 92.0% classification sensitivity. Elman's neural network's classification sensitivity was 99.00%. Else, they found that, with feedforward neural networks, the average accuracy was 86.6%, whereas for feedback, it was 98.8% [38].

In addition, Li et al. [39] presented a cascaded ML architecture for the purpose of DS prognosis. It is built on three steps: model ensemble via voting method, final judgment using logistic regression approach, and pre-judgment using isolation forest technique. The experimental findings demonstrate that, when assessed using various evaluation parameters, this framework outperforms some ML techniques in terms of performance on the maternal serum screening dataset. Li et al. showed that the category consisting of mother age, unconjugated estriol, human chorionic gonadotropin, and alpha-fetoprotein is the most recommended combination of input characteristics for DS screening. Their technique also has the ability to generate more accurate forecasts for feature-correlated and unbalanced data, offering a fresh and useful tool for the investigation of specific illnesses [39].

As well, A. Jamshidnezhad et al. [40] predicted DS using the first-trimester screening test by combining a genetic algorithm (GA) with an ANN. To investigate the suggested model, they gathered sample data from 381 expectant mothers who were directed to a for-profit screening reference laboratory between weeks 11 and 13 of gestation for an NT ultrasound and first-trimester screening test. In this research, A. Jamshidnezhad et al. suggested a feedforward neural network model and GA that defines both the model's topology and input parameters. They also used the sensitivity, specificity, and accuracy metrics to assess the performance of the developed model. The team demonstrated that the created model could reliably detect instances of DS with a specificity of 99.72%, sensitivity of 90.91%, and mean square error (MSE) of 0.61%, according to an average of ten studies. This study showed that by employing GA to optimize the neural network technique's structure, the accuracy of detecting DS could be raised using data from first-trimester screening tests [40].

Similarly, F. He et al. [41] developed an ML model based on a random forest. In total, 58,972 expectant mothers were incorporated into the model, comprising 49 cases of DS who had finished their screening in the second trimester. To confirm the predicted accuracy of the model, they also applied it to a different hospital dataset with 27,170 pregnant patients, 27 of whom were instances of DS. With the model dataset having a 5% false positive rate, they found that this ML model produced a DS detection rate of 66.7%. Furthermore, F. He et al. demonstrated that the ML model produced an 85.2% DS detection rate with a 5% false positive rate in the external verification dataset. This triple test for DS in second-trimester pregnancies has a false positive rate of 5% and a detection rate of approximately 60–65%. To conclude, the team compared the ML model to the current laboratory risk model, and they found that it enhances the DS detection rate while maintaining the same false positive rate, although the change is not statistically significant [41].

Yet, Sun et al. [42] created and validated a nomogram with fetal nuchal translucency thickness (NT) and ultrasonographic face indicators, to perform a trisomy 21 screening during the first trimester of gestation. Using archived two-dimensional mid-sagittal fetal profile images from 11 + 0 to 13 + 6 weeks' pregnancy in singleton pregnancies, they made the study as a retrospective case-control analysis. They automatically selected the discriminative markers using multivariable analysis and the approach known as the least absolute shrinkage and selection operator (LASSO). They also assessed the prediction effectiveness of trisomy 21 screening using the area under the receiver-operating-characteristic curve (AUC) using the LASSO model and a model using mother age and fetal NT. After that, they evaluated the nomogram's performance through the use of the calibration curve and C-index. As a result, the team found that the AUCs of the eight distinct ultrasonographic indicators that were part of the trisomy 21 screening LASSO model were 0.979 (95% CI, 0.966–0.993) within the validation collection, and 0.983 (95% CI, 0.971–0.994) in the training set. In the retest set, the LASSO model's AUC was 0.997 (95% CI, 0.990–1.000), demonstrating the model's strong resilience. In the sets used for training and validation, and when comparing the AUC of the LASSO model to the model based on fetal NT and maternal age, the difference was statistically significant ($p < 0.001$) [42].

Moreover, Xu et al. [29] applied NIPT to 856 patients at high-risk single pregnancies throughout the initial and middle stages of pregnancy in order to investigate the soft ultrasound marker (USM) in conjunction with NIPT in diagnosing fetal chromosomal abnormalities using ML and data mining techniques. To determine the effectiveness of USM, DS screening, and NIPT in identifying fetal chromosomal abnormalities, Xu et al. subjected all 856 individuals to amniocentesis and chromosome karyotype analysis. They found that 36 fetuses (4.21%) and 129 fetuses (15.07%) with two or more positive USM were among the 856 fetuses, and 96 fetuses (9.46%) have anomalies related to their chromosomes. In addition, they discovered that chromosome abnormalities were present in 36.11% of the group with multiple USM. It was greater than the 6.22% ($p < 0.01$) group without USM and the 19.38% ($p < 0.05$) group with only one USM. Then, when they combined the USM, Down's syndrome screening, and NIPT to diagnose fetal chromosomal abnormalities, the sensitivity, specificity, and accuracy were 96.72%, 98.45%, and 98.29% [29].

Finally, Chaplot et al. [43] studied the genetic disorder and genetic disorder sub-classes datasets for a variety of ML algorithms, including KNN, support vector machine, decision trees, Gaussian Naïve Bayes, logistic regression, gradient boosting, light gradient boosting classifier, random forest, extreme gradient boosting classifier, as well as CatBoost. They first pre-processed the dataset to look for NAN values, which are then visually shown in a number of categories to help with pattern analysis. These categories include genetic disorders, genetic disease sub-classes, five symptom samples, genes inherited from both parents, birth abnormalities, and more. After that, they chose features using a standardization approach, to which ML models were applied. Then, the team assessed the features using the accuracy, precision, recall, F1-score, loss, and root MSE. In addition, they produced the confusion matrix in order to calculate the values of true positive, false positive, and false negative for the classes selected from both datasets. As a result, they found that only the random forest, decision tree, LGBM classifier, XGB classifier, and CatBoost presented an accuracy of 99.9% for genetic disorder sub-classes. On the other hand, they found that for genetic disorder, the highest accuracy was calculated by decision tree, random forest, gradient boosting, LGBM classifier, XGB classifier, and CatBoost by 99.9% [43].

8.3.1.1 Discussion on the potential biases, limitations, and applicability of the results of machine learning approaches

Even if ML techniques have found great effectiveness in predicting and detecting down symptoms, they have also shown limitations in their use. For instance, the first study conducted by Yang et al. showed that the non-Z-score characteristics used for current SVM model training were not substantially skewed in the distributions of negatives and positives, which might be attributed to changes in sample composition [35]. In the case of Koivu et al., the use of multiple classification techniques using two real-world datasets was limited by the comparatively small number of data points [36], whereas Neocleous et al. demonstrated that the disadvantage of AI classification techniques is that it does not classify properly, in both phases, the euploid cases as no risk [37]. Else, Catic et al. found that among the main limitations of feedforward neural networks is the employment of just one health center and only one equipment during the study. Also, this study is limited by the number of "abnormal" samples [38]. While A. Jamshidnezhad et al. revealed that GA, and ANN needs specific equipment, and its downsides include the high cost, a delay in getting test results, and the difficulty of separating fetal DNA from the mother's blood [40]. On the other hand, F. He et al. highlighted that during the application of random forest algorithm, they did not consider the data of singleton pregnancies and natural conceptions, because there wasn't enough information available about multiple births and assisted reproduction [41]. In other ways, yet, Sun et al. indicated that the limitations of nomogram using fetal NT and ultrasonographic face indicators include the exclusion of fetuses with ineligible pictures and those without a nasal bone. This exclusion might have contributed to bias in their outcomes [42]. Nevertheless, Xu et al. found that the very small sample size is the primary source of restriction for the application of NIPT, and the soft USM technologies. Researchers must create, train, and test statistical ML models using greater datasets that contain a greater variety of demographic and clinical factors to completely explore the possibilities of these algorithms [29].

8.3.2 Deep learning approaches for Down syndrome diagnosis

Feng et al. [44] analyzed a recently released Illumina genotyping array and applied DL techniques to create reliable DS prediction/screening models. They used clinical genotyping insights from the Medical Center of Vanderbilt University to create chromosomal SNP maps. Next, using two combined convolutional neural network (CNN) models and a ten-layer CNN architecture, they combined two input chromosomal SNP maps. In addition to achieving an extremely reduced false positive and false negative rate, both of which are crucial for the screening and the prediction of the disease in healthcare practice, their CNN DS screening/prediction algorithm obtained an average accuracy of approximately 99.3%. Once compared to three traditional ML models, it performed superior across all evaluation metrics [44].

Equally, Selfiana et al. [45] employed clustering to handle the problem that frequently comes when classifying DS as high or low risk. In order to examine which of the two unsupervised algorithms can predict the best clustering, they were required to use DBSCAN and k-means clustering. Selfiana et al.

expected to see low distances between the clusters that are executed and noise-free clustering. In consequence, they found that, according to the data, clustering outperforms DBSCAN in terms of prediction accuracy for identifying DS, a mixture of multi-branch CNN and feature rearrangement for predicting DS even in the presence of slightly elevated noise, which DBSCAN is the only method capable of eliminating. This is indicated by k-means clustering [45].

Furthermore, using CNNs and face photos, Qin et al. [46] created a technique for identifying DS that quantified the binary classification issue of separating people with DS from healthy ones using unconstrained two-dimensional images. Based on a large-scale face identification dataset (10,562 participants), they first created a generic facial recognition network. Next, they trained (70%) and evaluated (30%) a database of 148 photos with DS and 257 images with healthy subjects that were picked from public sources. At the last testing, the deep CNN identified people with DS with specificity of 97.40%, accuracy of 95.87%, and recall of 93.18% [46].

According to estimates by A.R. Porras et al. [47], the Democratic Republic of the Congo has one of the highest incidences of birth abnormalities worldwide. Therefore, the A.R. Porras et al. team set out to find out if people of Congolese origin could reliably be diagnosed with DS using smartphone-based face analysis technologies. Before receiving technological training, they used inexpensive genomic applications, which are widely accessible in low-resource nations and do not require sophisticated technical knowledge to validate the existence of trisomy 21. When compared to earlier software developed on non-Congolese persons, their program, trained on 132 Congolese participants, performed much better (91.67% accuracy, 95.45% sensitivity, and 87.88% specificity; $p < 5\%$). They also present a list of the most distinctive facial traits associated with DS, along with their respective distributions throughout the Congolese population. As a consequence, the team confirmed that, when combined, these technologies provide the local people with an accurate and affordable way to diagnose DS [47].

Besides, Ji et al. [48] investigated the medical use of AI model for detecting fetal anomalies in the first trimester, as well as the validity of the model for evaluating fetal facial profile indicators. They utilized two-dimensional mid-sagittal fetal profile pictures obtained at 11-13+6 weeks of gestation during singleton pregnancies in this retrospective investigation. Else, they created an AI model using semantic segmentation and landmark localization to quantify the chosen markers and assess the diagnostic value for prenatal anomalies. Outcomes, they observed a total of 37 anomalous fetuses, encompassing 18 with trisomy 21, 12 with CLP, and 7 with trisomy 18, out of 2372 normal newborns. Of these, Ji et al. used all the babies with anomalies and the remaining 500 normal fetuses for clinical testing, while they utilized the 1872 normal fetuses for training and validating AI models. As a result, the team found that the 500 normal fetuses' ICCs (95% CI) for the IFA, MNM angle, FMA, FS distance, and PL distance were 0.812 (0.780–0.840),

0.760 (0.720–0.795), 0.766 (0.727–0.800), 0.807 (0.775–0.836), and 0.798 (0.764–0.828) for the AI and manual measurements, respectively. Therefore, they demonstrated that trisomy 21 and 18 were clinically and substantially recognized by IFA; their respective areas under the ROC curve (AUC) were 0.686 (95% CI, 0.585–0.788) and 0.729 (95% CI, 0.621–0.837). With an AUC of 0.904 (95% CI, 0.842–0.966), FMA predicted trisomy 18 accurately. The AUCs for MNM angle and FS distance in CLP were 0.738 (95% CI, 0.573–0.902) and 0.677 (95% CI, 0.494–0.859), respectively, indicating strong predictive value [48].

Too, Phung et al. [49] suggest utilizing a feature rearrangement technique in conjunction with a multi-branch CNN model to enhance the accuracy of DS prediction using prenatal screening dataset. To properly organize the features for the CNN algorithm, they proposed a feature rearrangement method that makes use of Pearson correlation testing and feature grouping. They observed that the studies achieved good outcomes with 89.69% F1-score, 90.23% recall, and 0.93.14% balanced accuracy, despite the severely missing and unbalanced data [49].

8.3.2.1 Discussion on the potential biases, limitations, and applicability of the results of deep learning approaches

Similar to ML, researchers have faced several challenges when applying DL. For instance, Feng et al. demonstrated that the use of two combined CNN models and ten-layer CNN architecture was restricted by the computing expenses associated with discovering definite characteristics for subset characterization and optimization [44]. In addition, A.R. Porras et al. observed that smartphone-based face analysis technologies could be impacted by the dataset's inadequate matching [47]. Finally, Ji et al. study with the use of two-dimensional mid-sagittal fetal profile pictures presented three limitations. The restricted number of fetal abnormalities notices may have influenced the precision of the model in identifying particular kinds of abnormalities; second, the greater part of mid-sagittal section images were easily investigated; third, the primary limitation of this study is its retrospective design [48].

8.4 DISCUSSION

Despite the widespread use of laboratory tests as diagnostic tools, data suggest that the use of AI could aid in the accurate detection of DS. As far as we know, this systematic review is the first to examine the use of AI for predicting DS. In expansion to the over said detailed work, the comparative investigation outlined in Table 8.1 grandstands the point by point data such as sort of dataset, procedures, and the anticipated results with respect to the work done by the researchers, which in return made a difference to the author to hunt for the leading strategy for identifying or diagnosing DS.

Table 8.1 AI models test results with datasets

Reference	*AI technique*	*Results*
[35]	-SVM discriminating model	-Accuracy: 100%
[36]	-Multiple classification techniques using two real-world datasets	-Detection rate: 78% -False positive rate: 1%
[37]	-AI classification techniques	-Sensitivity: 97.1% -Specificity: 99.5%
[38]	-Feedforward neural networks	-Accuracy: 86.6%
[39]	-Final judgment using logistic regression approach -Pre-judgment using isolation forest technique	-The most recommended combination: The group of alpha-fetoprotein, human chorionic gonadotropin, unconjugated estriol, and maternal age
[40]	-Genetic algorithm (GA) -Artificial neural network (ANN)	Specificity: 99.72% -Sensitivity: 90.91% -MSE: 0.61%
[41]	-Random forest	-DS detection rate: 85.2% -False positive rate 5%
[42]	-Nomogram based on fetal nuchal translucency thickness (NT) -Ultrasonographic face indicators	In the validation set: -AUC: 0.979 (95% CI, 0.966–0.993) In the training set: -AUC: 0.983 (95% CI, 0.971–0.994)
[29]	-NIPT -The soft ultrasound marker (USM)	-Sensitivity: 96.72% -Specificity: 98.45% -Accuracy: 98.29%
[45]	-DBSCAN -k-means clustering	k-means that clustering is better than DBSCAN where the prediction level of clustering is better for determining Down syndrome
[43]	-Support vector machines -Gaussian Naïve Bayes -KNN -Decision trees -Gradient boosting -Logistic regression -Light gradient boosting classifier -Random forest -Extreme gradient boosting classifier -CatBoost	-Genetic disorder accuracy: 99.9% (by decision tree, random forest, gradient boosting, LGBM classifier, XGB classifier, and CatBoost) -Genetic disorder sub-classes accuracy: 99.9% (by random forest, decision tree, LGBM classifier, and CatBoost)
[44]	-Two combined CNN models -Ten-layer convolutional neural network (CNN) architecture	-Accuracy: 99.3%
[46]	-Deep CNNs -Face photos	-Accuracy: 95.87% -Recall: 93.18% -Specificity: 97.40%

(*Continued*)

Table 8.1 (Continued)

Reference	*AI technique*	*Results*
[47]	-Smartphone-based face analysis technologies	-Accuracy: 91.67% -Sensitivity: 95.45% -Specificity: 87.88%
[48]	-Two-dimensional mid-sagittal fetal profile pictures	-A good consistency of fetal facial profile marker measurements between the AI and manual measurement during the first trimester.
[49]	-Feature rearrangement technique -Multi-branch CNN model	-Recall: 0.9023 -F1-score: 0.8969 -Balanced accuracy: 0.9314

According to these insights, it can be observed that the use of AI is promising to predict DS in the earlier stage of pregnancy. These studies indicated that detection of the most important signs of this disease is possible now and easier with the use of ML and DL technologies. As an example of case studies, the work carried out by A.R. Porras et al. in 2021 applied in the Democratic Republic of the Congo that known as one of the highest incidences of birth abnormalities worldwide. As a result, the team confirmed that, smartphone-based face analysis technologies provide the local people with an accurate and economical approach to detect DS [47]. Besides, ML algorithms can be integrated into the cell-free prenatal screening for the diagnosis of DS. This will offer a paradigm shift with important incentives and significant improvements. Concerns about pointless interventions may be allayed by the possibility of decreased false positives and enhanced early identification. The adaptive nature of machine learning guarantees continuous enhancement participating to the development of more accurate and valid prediction of prenatal DS. These ML algorithms allow for the study of enormous datasets containing clinical and genetic data, potentially uncovering narrow markers and patterns boosting the detection accuracy over classical techniques [50].

Even if AI offers promising advancements in prenatal screenings for DS, its implementation introduces a host of ethical considerations that demand thorough examination. False positives may trigger unnecessary anxiety and annoying procedures for expectant parents, while false negatives could result in missed opportunities for timely interventions and support. Alongside clinical accuracy, a thorough understanding of the ethical, social, and psychological consequences of AI-powered prenatal tests for DS is essential. This includes considerations of reproductive autonomy, the rights and dignity of individuals with DS, societal perceptions, and the emotional impact on expectant parents. By addressing these multifaceted dimensions, we can ensure a balanced perspective that informs ethical decision-making and fosters compassionate care in prenatal screening practices.

8.5 CONCLUSION

DS, resulting from trisomy of chromosome 21, is a well-known chromosomal illness in humans. Several distinct teams of scientists have used knowledge exploration approach in a range of fields to accurately predict having a child with DS. To facilitate the diagnostic process, scientists explored AI for DS prediction in the first- and second-trimester antenatal checking. The AI technique is equivalent to the conventional approach and could be presented as an alternate technique for predicting DS during antenatal testing. AI, as a practical and effective method for primary diagnosis for fetal trisomy, allows for timely medical assessment of fetal outcome and supports female reproductive wellness.

As perspective, we are looking forward to investigate the integration of multimodal data, including genomic, clinical, and imaging data, to enhance the performance of predictive algorithms. Additionally, we intend to conduct a large-scale genomic analysis to identify novel genetic biomarkers and molecular signatures associated with DS.

REFERENCES

1. Häggsgård C, Nilsson C, Teleman P, Rubertsson C, Edqvist M. Women's experiences of the second stage of labour. *Women Birth*. 35(5), e464–e470 (2022).
2. Zietlow A-L, Nonnenmacher N, Reck C, Ditzen B, Müller M. Emotional stress during pregnancy – Associations with maternal anxiety disorders, infant cortisol reactivity, and mother–child interaction at pre-school age. *Front Psychol*. 10, 2179 (2019).
3. Aksoy S. Antenatal screening and its possible meaning from unborn baby's perspective. *BMC Med Ethics*. 2(1), 3 (2001).
4. Crombag NMTH, Boeije H, Iedema-Kuiper R, Schielen PCJI, Visser GHA, Bensing JM. Reasons for accepting or declining Down syndrome screening in Dutch prospective mothers within the context of national policy and healthcare system characteristics: A qualitative study. *BMC Pregnancy Childbirth*. 16(1), 121 (2016).
5. Yao Y, Liao Y, Han M, Li S-L, Luo J, Zhang B. Two kinds of common prenatal screening tests for Down's syndrome: A systematic review and meta-analysis. *Sci Rep*. 6(1), 18866 (2016).
6. Antonarakis SE, Skotko BG, Rafii MS, *et al*. Down syndrome. *Nat Rev Dis Primers*. 6(1), 9 (2020).
7. MacLennan S. Down's syndrome. *InnovAiT*. 13(1), 47–52 (2020).
8. Zhao Q, Rosenbaum K, Sze R, Zand D, Summar M, Linguraru MG. *Down Syndrome Detection from Facial Photographs Using Machine Learning Techniques [Internet]*. Novak CL, Aylward S (Eds.), Lake Buena Vista (Orlando Area), Florida, USA, 867003 (2013) [cited 2024 Jan 12]. Available from: http://proceedings.spiedigitallibrary.org/proceeding.aspx?doi=10.1117/12.2007267.
9. Vuorenlehto L, Hinnelä K, Äyräs O, Ulander V-M, Louhiala P, Kaijomaa M. Women's experiences of counselling in cases of a screen-positive prenatal screening result. *PLoS ONE*. 16(3), e0247164 (2021).

10. Phadke SR, Puri RD, Ranganath P. Prenatal screening for genetic disorders: Suggested guidelines for the Indian Scenario. *Indian J Med Res.* 146(6), 689–699 (2017).
11. Wu X, Sahoo D, Hoi SCH. Recent advances in deep learning for object detection. *Neurocomputing.* 396, 39–64 (2020).
12. Shang W, Wan Y, Chen J, Du Y, Huang J. Introducing the non-invasive prenatal testing for detection of Down syndrome in China: A cost-effectiveness analysis. *BMJ Open.* 11(7), e046582 (2021).
13. Allyse M, Minear M, Rote M, *et al.* Non-invasive prenatal testing: A review of international implementation and challenges. *IJWH.* 113 (2015).
14. Norton ME, Jacobsson B, Swamy GK, *et al.* Cell-free DNA analysis for noninvasive examination of trisomy. *N Engl J Med.* 372(17), 1589–1597 (2015).
15. Kater-Kuipers A, Bunnik EM, de Beaufort ID, Galjaard RJH. Limits to the scope of non-invasive prenatal testing (NIPT): An Analysis of the international ethical framework for prenatal screening And An interview study with Dutch professionals. *BMC Pregnancy Childbirth.* 18(1), 409 (2018).
16. Lutgendorf MA, Stoll KA, Knutzen DM, Foglia LM. Noninvasive prenatal testing: Limitations and unanswered questions. *Genet Med.* 16(4), 281–285 (2014).
17. Qadri AM, Raza A, Eid F, Abualigah L. A novel transfer learning-based model for diagnosing malaria from parasitized and uninfected red blood cell images. *Decis Anal J.* 9, 100352 (2023).
18. Jha K, Doshi A, Patel P, Shah M. A comprehensive review on automation in agriculture using artificial intelligence. *Artif Intell Agric.* 2, 1–12 (2019).
19. Wingfield LR, Ceresa C, Thorogood S, Fleuriot J, Knight S. Using artificial intelligence for predicting survival of individual grafts in liver transplantation: A systematic review. *Liver Transpl.* 26(7), 922–934 (2020).
20. Begum A, Kumar R. Review of chronic inflammation and long term effects on health using machine learning algorithms. *IJCSE.* 9(6), 64–71 (2021).
21. Akhtar M, Moridpour S. A review of traffic congestion prediction using artificial intelligence. *J Adv Transp.* 2021, 1–18 (2021).
22. Rebala G, Ravi A, Churiwala S. Machine Learning Definition and Basics [Internet]. In: *An Introduction to Machine Learning*, Springer International Publishing, Cham, 1–17 (2019). [cited 2023 Dec 19]. Available from: http://link.springer.com/10.1007/978-3-030-15729-6_1.
23. Park DJ, Park MW, Lee H, Kim Y-J, Kim Y, Park YH. Development of machine learning model for diagnostic disease prediction based on laboratory tests. *Sci Rep.* 11(1), 7567 (2021).
24. Santangelo OE, Gentile V, Pizzo S, Giordano D, Cedrone F. Machine learning and prediction of infectious diseases: A systematic review. *MAKE.* 5(1), 175–198 (2023).
25. Potnis N, Tiple B. Machine learning techniques for disease prediction. *ITM Web Conf..* 57, 01004 (2023).
26. Alanazi R. Identification and prediction of chronic diseases using machine learning approach. *J Healthc Eng.* 2022, 1–9 (2022).
27. Neagos D, Cretu R, Sfetea RC, Bohiltea LC. The importance of screening and prenatal diagnosis in the identification of the numerical chromosomal abnormalities. *Maedica (Bucur).* 6(3), 179–184 (2011).
28. Abedalthagafi M, Bawazeer S, Fawaz RI, Heritage AM, Alajaji NM, Faqeih E. Non-invasive prenatal testing: A revolutionary journey in prenatal testing. *Front. Med.* 10, 1265090 (2023).
29. Xu X, Wang L, Cheng X, *et al.* Machine learning-based evaluation of application value of the USM combined with NIPT in the diagnosis of fetal chromosomal abnormalities. *MBE.* 19(4), 4260–4276 (2022).

30. Gori C, Cocchi G, Corvaglia LT, *et al.* Down syndrome: How to communicate the diagnosis. *Ital J Pediatr.* 49(1), 18 (2023).
31. Ho L. Current status of the early childhood developmental intervention ecosystem in Singapore. *SMEDJ.* 62(1 Suppl), S43–S52 (2021).
32. Paterick TE, Patel N, Tajik AJ, Chandrasekaran K. Improving health outcomes through patient education and partnerships with patients. *Bayl Univ Med Cent Proc.* 30(1), 112–113 (2017).
33. Wills J. Health literacy: New packaging for health education or radical movement? *Int J Public Health.* 54(1), 3–4 (2009).
34. Nutbeam D. The evolving concept of health literacy. *Soc Sci Med.* 67(12), 2072–2078 (2008).
35. Yang J, Ding X, Zhu W. Improving the calling of non-invasive prenatal testing on 13-/18-/21-trisomy by support vector machine discrimination. Bioinformatics (2018). Available from: http://biorxiv.org/lookup/doi/10.1101/216689.
36. Koivu A, Korpimäki T, Kivelä P, Pahikkala T, Sairanen M. Evaluation of machine learning algorithms for improved risk assessment for Down's syndrome. *Comput Biol Med.* 98, 1–7 (2018).
37. Neocleous AC, Syngelaki A, Nicolaides KH, Schizas CN. Two-stage approach for risk estimation of fetal trisomy 21 and other aneuploidies using computational intelligence systems. *Ultrasound Obstet Gyne.* 51(4), 503–508 (2018).
38. Catic A, Gurbeta L, Kurtovic-Kozaric A, Mehmedbasic S, Badnjevic A. Application of neural networks for classification of Patau, Edwards, Down, Turner and Klinefelter syndrome based on first trimester maternal serum screening data, ultrasonographic findings and patient demographics. *BMC Med Genomics.* 11(1), 19 (2018).
39. Li L, Liu W, Zhang H, Jiang Y, Hu X, Liu R. Down syndrome prediction using a cascaded machine learning framework designed for imbalanced and feature-correlated data. *IEEE Access.* 7, 97582–97593 (2019).
40. Jamshidnezhad A, Hosseini SM, Mohammadi-Asl J, Mahmudi M. An intelligent prenatal screening system for the prediction of Trisomy-21. *Inf Med Unlocked.* 24, 100625 (2021).
41. He F, Lin B, Mou K, Jin L, Liu J. A machine learning model for the prediction of down syndrome in second trimester antenatal screening. *Clin Chim Acta.* 521, 206–211 (2021).
42. Sun Y, Zhang L, Dong D, *et al.* Application of an individualized nomogram in first-trimester screening for trisomy 21. *Ultrasound Obstet Gyne.* 58(1), 56–66 (2021).
43. Chaplot N, Pandey D, Kumar Y, Sisodia PS. A comprehensive analysis of artificial intelligence techniques for the prediction and prognosis of genetic disorders using various gene disorders. *Arch Comput Methods Eng.* 30(5), 3301–3323 (2023).
44. Feng B, Samuels DC, Hoskins W, *et al.* Down syndrome prediction/screening model based on deep learning and Illumina genotyping array [Internet]. In: *2017 IEEE International Conference on Bioinformatics and Biomedicine (BIBM)*, IEEE, Kansas City, MO, 347–352 (2017) [cited 2024 Jan 10]. Available from: http://ieeexplore.ieee.org/document/8217674/.
45. Selfiana R, Sudarmilah E, Putri DAP. Comparison of K-means and DBSCAN for prediction determination of Down syndrome using prenatal test data [Internet]. Chennai, India, 040014 (2023) [cited 2024 Jan 10]. Available from:http://aip.scitation.org/doi/abs/10.1063/5.0141770.
46. Qin B, Liang L, Wu J, Quan Q, Wang Z, Li D. Automatic identification of Down syndrome using facial images with deep convolutional neural network. *Diagnostics.* 10(7), 487 (2020).

47. Porras AR, Bramble MS, Mosema Be Amoti K, *et al.* Facial analysis technology for the detection of Down syndrome in the Democratic Republic of the Congo. *Eur J Med Genet.* 64(9), 104267 (2021).
48. Ji C, Liu K, Yang X, *et al.* A novel artificial intelligence model for fetal facial profile marker measurement during the first trimester. *BMC Pregnancy Childbirth.* 23(1), 718 (2023).
49. Phung NH, Nguyen CT, Tran TK, *et al.* A Combination of Multi-Branch CNN and Feature Rearrangement for Down Syndrome Prediction [Internet]. In: *2023 International Conference on Artificial Intelligence in Information and Communication (ICAIIC)*, IEEE, Bali, Indonesia, 1–6 (2023) [cited 2024 Jan 10]. Available from: https://ieeexplore.ieee.org/document/10067118/.
50. Baldo F, Piovesan A, Rakvin M, *et al.* Machine learning based analysis for intellectual disability in Down syndrome. *Heliyon.* 9(9), e19444 (2023).

Chapter 9

Blockchain technology in healthcare

Improving security, privacy, and secure transactions in patient-centered systems

Sanae Seidi and Abderrahim Abdellaoui

9.1 INTRODUCTION

The blockchain hype has reached a height in recent years, largely fueled by the remarkable success of Bitcoin. The potential for growth in blockchain-based applications extends across various sectors, including healthcare [1]. Foreseeably, there is an upward trajectory in the adoption of blockchain technology across industries such as healthcare [2, 3], finance [4], commerce [5], Internet of Things (IoT) [6], risk management systems [7], government transactions [8], and royalty payments [9]. Notably, blockchain platforms are perceived as secure [10] due to their inherent feature of tracking every participant's actions meticulously. Any alterations to the blocks become immensely challenging to detect without computational prowess [11].

Blockchain technology facilitates "peer-to-peer" online transactions between individuals. It is important to note that cryptocurrencies induce users at both ends to do services in a decentralized, automated, and trustless way. Blockchains operate as open, decentralized peer-to-peer networks keeping in view the appending-only and trustless public ledger. The trustlessness of blockchain technology stems from its decentralized and transparent nature, where every system operation occurs autonomously and transparently. This trustless system is established as all peers become ledger members, forming a distributed ledger that records and logs every transaction. The rules governing this decentralized ledger are predetermined and remain consistent across all wallet users, ensuring each transaction has multiple sources of legitimacy. Utilizing a distributed network of nodes, transactions are verified through a mining process. While blockchains impose adherence to established rules and distribute trust among multiple parties involved in the system, they do not entirely eliminate it [12].

Our healthcare needs are catered to by healthcare systems, and with the aging population, there arises a greater demand for productivity [13]. Particularly highlighted by the COVID-19 pandemic, real-time information becomes paramount, emphasizing the importance of patient health data such as temperature or symptoms [14, 15]. As more individuals develop chronic illnesses due to aging demographics, there emerges a pressing need for enhanced patient care, thereby elevating awareness of health-related

DOI: 10.1201/9781003478676-9

issues [16]. The appearance of the IoT and the Internet of Medical Things (IoMT) are the most recent tech advancements in the healthcare system. Nevertheless, the concentrated nature of the processing, storing, and utilizing it could have its limitations such as data manipulation, disbelief, privacy issues, and single-point failures. Safeguarding the security of all internet-enabled devices remains a significant concern in the increasingly technology-driven healthcare landscape [17]. The integration of blockchain technology with healthcare yields improved security, decentralized processing, and storage. This integration holds promise, granting medical staff and other stakeholders in healthcare institutions enhanced access to verified and validated records, thereby improving healthcare assistance [18, 19], interoperability [20], efficient storage of health records [19, 21], and security systems [22].

The contributions of the work can be summarized as follows:

1. Fundamentals of Blockchain: Offering a detailed understanding of the workings, operations, architecture, limits, and motivations of blockchain technology in the healthcare industry.
2. Blockchain Overview in Healthcare: Providing insights into the features and various use cases of blockchain technology within the healthcare industry.
3. Strategic Analysis of Blockchain Adoption in Healthcare: Exploring the strategic dimension of blockchain adoption in healthcare through a comprehensive SWOT (Strengths, Weaknesses, Opportunities, and Threats) Analysis.
4. Comparative Analysis of Healthcare Applications: Examining existing applications of blockchain technology in healthcare, with a focus on their functionalities and use cases.
5. Discussion of Comparative Analysis: Engaging in a thorough discussion based on the accumulated results from each axis of analysis.

In this chapter, we start by giving a general introduction to blockchain technology, including its uses, characteristics, kinds, consensus processes, constraints, and driving forces. We next concentrate on its uses in the healthcare industry before doing a SWOT-based strategic analysis. We then present a comparative analysis of many blockchain-based healthcare systems. The ramifications of our findings are examined in our discussion section, which also provides recommendations for further study and useful information for healthcare stakeholders. All things considered, our research advances knowledge about blockchain's application in healthcare and directs decision-making in this developing area.

9.2 BLOCKCHAIN: OVERVIEW

Blockchain has been the focus of rather extensive research since its 2008 inception. For example, research has been done to identify blockchain applications (such as smart contracts and identity verification) [8, 9, 23]. In this part, we'll provide a quick rundown of blockchain, including its features and uses.

9.2.1 Blockchain applications

Blockchain boasts a broad spectrum of applications spanning various industries, leveraging its inherent characteristics and technological capabilities. It's important to distinguish between applications and features: applications refer to processes where blockchain technology can be employed to fulfill specific requirements, while features pertain to the inherent attributes of the system itself.

9.2.1.1 Smart contracts

Smart contracts represent a significant application of blockchain technology, allowing users to create legal documents directly within the blockchain system. These contracts, which act as autonomous agents, are stored within the blockchain and are utilized to execute various legal services. Upon encoding transactions, these smart contracts generate contracts and other legal documents.

The scripts constituting these smart contracts are stored on the blockchain and assigned distinct addresses, facilitating their location and validation. By enabling decentralized communication between participants and ensuring fair trade, smart contracts streamline the transfer of assets and deeds. Users can create their own paperwork without relying on notaries or legal advisors, thereby eliminating several time- and cost-related barriers [24].

9.2.1.2 Fraud detection

Blockchain technology finds another application in fraud detection, which involves scrutinizing documents or data systems for any signs of manipulation. In the banking sector, for instance, blockchain aids in preventing fact-based fraud, such as abusive loan application behavior. Moreover, it helps in curbing fraudulent activities like ballot stuffing and badmouthing, which contribute to fake reviews on online platforms [25].

Furthermore, blockchain technology extends its utility to the realm of crowdfunding, a relatively new concept. Crowdfunding involves raising equity for a company by selling stocks to numerous investors. Blockchain accelerates, streamlines, and secures the transfer of crowdfunding equity, facilitating peer-to-peer transactions between investors and entrepreneurs. Additionally, it serves as a cost-effective platform for stock and share registration. Blockchain also plays a pivotal role in preventing investment fraud, alerting authorities about market conditions, and establishing a shareholder voting system for corporate governance [26].

9.2.1.3 Identity verification

Outside of the healthcare industry, various internet companies have adopted alternative methods for identity verification. Modern times are when businesses and governments use identification cards like passports and fingerprints. In contrast to the standardized process where the government is responsible for the identification and verification of the paper documents, blockchain technology provides an innovative approach to identification verification in this sector. Blockchain technology can confirm a user's identification independently of these governments. one use of this is the notarization of birth certificates, business transactions, and marriages using blockchain technology. a person could demonstrate their existence at a specific location and time by using the distributed ledger feature of the blockchain, which is validated by a number of others [23].

9.2.2 Blockchain features

Numerous properties of blockchain technology make it useful for the healthcare sector. These functions are built into the system and are applicable to many different kinds of systems and sectors. This section will address security, authentication, and decentralized storage in particular (Figure 9.1).

9.2.2.1 Decentralized storage

Decentralized storage serves as a foundational element of blockchain technology, enhancing data security and authentication within the system [27, 28]. By leveraging the blockchain ledger, decentralized storage disperses record storage from a single centralized server to multiple servers. This decentralized approach facilitates quicker access to medical data, empowers patient agency, promotes system interoperability, and augments the quantity and quality of data available for medical research [27].

For instance, cloud and IoT providers can utilize blockchain technology to transmit data in a decentralized, private, and secure manner [27]. Immutability, integrity, and data security are paramount features of blockchain systems. Through its private, secure, decentralized ledger, blockchain

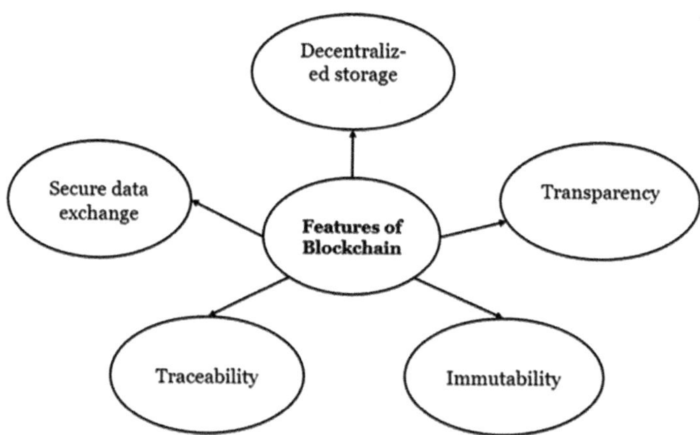

Figure 9.1 Features of blockchain.

stores data across multiple computers rather than a single source, thereby enhancing security. Utilizing a wide-area database strategy, noting down the transaction in every node of the whole network, a distributed backup will be produced automatically to make sure that the historical accuracy of the transaction is preserved [8, 9]. This redundancy achieves immutability, integrity, and security so that any interference by malicious agents would result in an impairment in data across every system in the network rather than this data system being separated from other systems and compromising the whole network.

The security issue, integrity, and individuals' privacy do not miss out in the case of smart devices, the public cloud, and the whole IoT ecosystem. In theory, blockchain technology as well as other tools can prevent the vulnerabilities of IoT devices because the distributed ledger technology makes it difficult for hackers to penetrate the network [28, 29]. By employing a decentralized blockchain infrastructure, cloud services can extend to edge hosts such as smart or IoT devices while maintaining additional security and control [29]. Furthermore, blockchain enables major data institutions to securely communicate medical data by providing cloud-based servers for data provenance, auditing, and control. Access control and smart contracts can be implemented to safeguard cloud-based infrastructures [30].

Additionally, blockchain technology can enhance client-server systems by introducing auditable security enhancements and enabling servers to respond to client requests more effectively [31].

9.2.2.2 *Transparency*

It is among the most intriguing features of blockchain technology. Every participant has access to see every transaction that is completed [32]. Data integrity is guaranteed as a result. Since they are digitally confirmed with the agreement of all participants, they cannot be compromised. Any type of alteration or modification requires the consent of each member, therefore data deletion or tampering is impossible without everyone's agreement. Nevertheless, there are problems with putting transparency into practice. Blockchain's high cost and energy usage could hinder its adoption. Furthermore, the integrity of the ledger could be in danger if it is not possible to confirm the validity of smart contracts and transaction verification. Users of blockchain technology must also be able to view the change policy [33].

The infrastructure that makes use of this ledger is necessary to tackle several problems. Nevertheless, there is no legal obligation to define this infrastructure, which means that there is a dearth of information available on the firmware or software. Consequently, each of these issues needs to be considered as the infrastructure is developed in order for blockchain technology to be applied in democratic contexts and effectively promote transparency.

Hospitals already electronically save patient data as an Electronic Health Record (EHR) in the current healthcare system. Openness and trust between organizations, however, have never been without issues. This issue can be resolved by the blockchain, which provides an immutable record with controlled data access. Once data is collected, it cannot be removed, making it constantly accessible for on-demand review at a later time. The degree to which all parties involved in a supply chain are able to access and share information about the product is known as transparency [34].

9.2.2.3 Immutability

Blockchain also has the critical virtue of immutability [35]. It is defined as the persistent state of a blockchain ledger that does not change while the blockchain remains persistent. The fact that blockchains are distributed, decentralized systems that function according to a consensus reached by the nodes that store the copy of the data must also be understood. The uniqueness of the data will be preserved, according to this agreement. Centralized databases require third-party assurance for security because they are vulnerable to hacking and security flaws. Blockchain updates ledgers whenever a transaction occurs [36]. The blockchain's participants exchange ledgers. Once a transaction is loaded onto the blockchain, it cannot be undone. In blockchain technology, transactions are gathered into blocks linked to each other using unique hash keys. Each block's transactions are verified, and if modified, a new hash key is created, allowing for easy tracking across other blocks. Ledgers on blockchain are constantly updated, keeping them synchronized and available to all users. With a vast network, manipulating ledgers becomes incredibly challenging, even if someone controls a majority of the network (51%) [37].

9.2.2.4 Traceability

Traceability stands out as another notable characteristic within blockchain technology, enabling transactions to be seamlessly tracked across the network in an immutable manner [38]. With every event visible to all peers, it becomes effortless to monitor the responsible parties for each occurrence, the timing of events, and the authorization of transaction execution. This feature inherently encourages users to take accountability for their actions and uphold honesty in all interactions. While the recent surge in blockchain adoption has prompted many businesses to offer blockchain-enabled traceability solutions, the overall quality of data collected and exchanged has not seen significant improvement. Consequently, questions have arisen regarding the effective utilization of blockchain technology in supply chain scenarios. However, each new transaction added to a block remains immutable and is accompanied by a timestamped hash, facilitating seamless data tracing and ensuring the unalterability of information within the block. Therefore, every business registering specific data or transactions on the blockchain must uphold trustworthiness to ensure the authenticity and traceability of the recorded information.

The supply chain for healthcare is an intricate web of interconnected businesses that includes manufacturers, distributors, pharmacies, hospitals, patients, and suppliers of raw materials. Drug traceability, also known as tracking and tracing, is becoming increasingly necessary and is required in some nations. It is challenging to trace supply throughout this network due to a number of issues, including centralized management, contradictory stakeholder behavior, and a lack of information [39]. The traceability of drugs within the pharmaceutical supply chain has become one of the most important factors today considering its role in creating a reliable platform for performing the authentication process and also tracking the chain of custody of the product along the way in the supply chain.

9.2.2.5 *Secure data exchange*

Blockchain does away with the requirement for an administrator or third party to ensure the security of data exchange. Blockchain ensures consistency and reliability through a decentralized consensus-based access mechanism [40]. Blockchain technology uses public-key cryptography to secure and thwart transaction fraud. The identity network security system guards against fraud and data theft while permitting authorized access to digital signatures. Secure, reliable, compliant, auditable, and practical data sharing is essential for a successful data exchange. There are several intriguing ways that blockchain technology can make it easier, more transparent, and reliable for participants in the data ecosystem to transmit data. Blockchain ensures transparency, privacy, interoperability, and security by maintaining data's immutability, auditability, and single version of the truth. The chapter expounded a Blockchain-based framework named DASS-CARE which is a decentralized, transparent, scalable system for storing and accessing medical data [41]. This kind of architecture is capable of drastically shortening the time of procedures and the need to get patient info while the data remains linked with its security, integrity, and confidentiality. Furthermore, the document presented the security and information exchange strategy which was based on a hospital's single private chain [42].

Numerous security needs, like transparency, decentralization, and resilience to tampering, are satisfied by this architecture. Physicians have the ability to store medical records or access past patient information through a private, secure system [43]. Additionally, a method for matching symptoms between patients is offered. It makes it possible for patients with similar symptoms to authenticate one another and make a session key in case of illness later.

9.2.3 Blockchain type

Generally speaking, there are several kinds of blockchains based on the data that is handled, whether it is accessible, and what the user may do. Among them are public permissionless, consortium (public permission), and private.

The public has visibility and access to all of the data on the public permissionless (also referred to as just public) blockchain. However, in order to protect a participant's anonymity, some portions of the blockchain might be encrypted [44].

Anyone can join a public permissionless blockchain without requiring permission and can function as either a simple node or a miner (node). Typically, these kinds of blockchains are offered financial incentives, such as those found in cryptocurrency networks. A few instances of such a blockchain are Litecoin, Ethereum, and Bitcoin [45].

The distributed consensus process on the consortium-type blockchain is limited to a specific set of nodes [44].

It might be applied in a single industry or multiple. A consortium blockchain is partially centralized and available to the public for limited use when it is developed inside a specific industry, such as the banking sector. In contrast, a consortium comprising various industries such as insurance, finance, and government is made available to the general public, albeit with a somewhat centralized trust already in place.

A private blockchain restricts network access to selected nodes. As a result, it is a distributed but centralized network [44].

Permissioned networks, such as private blockchains, allow administrators to restrict which nodes are able to carry out transactions, carry out smart contracts, or mine. They are overseen by a single, reliable organization. It is employed for personal gain. Examples of blockchain technologies that only support private blockchain networks are Hyperledger Fabric [46] and Ripple [47] (Table 9.1).

Table 9.1 Comparison between different types of blockchain

Feature	*Public blockchain*	*Private blockchain*	*Consortium blockchain*
Consensus mechanism	All users	Single entity	Selected set of nodes
Read access	Community	Community or restricted	Community or restricted
Performance	Low	High	High
Centralization	No	Yes	Partial
Consensus process	Permissionless	Permissioned	Permissioned
Data immutability	Almost impossible	Vulnerable to tampering	Vulnerable to tampering
Scalability	Limited	High	High
Confidentiality level	Low	High	Medium
Participant control	Decentralized	Centralized	Partially centralized

9.2.4 Consensus

A decentralized peer-to-peer system can make choices without an official figure by using the consensus mechanism. The purpose of this method is to guarantee the veracity and honesty of records. Three different kinds of consensus processes exist:

- **Proof of Stake:** In this method, the validator, who also creates new blocks, is selected at random according to how much stakes they pledge to the network. The likelihood of becoming a validator increases with the quantity of stakes. Blockchain technologies that use the Proof of Stake consensus are Cardano's Ouroboros and EOS [48].
- **Proof of Work (PoW):** Transaction data in this system is kept in blocks linked to an intricate mathematical puzzle. Only when a problem is solved is such data verified. This procedure, called "mining," is carried out by powerful computers. The first miner to solve the puzzle will receive payment in Bitcoin. A PoW technique is used by cryptocurrencies like Ethereum and Bitcoin [48].
- **Proof of Authority** is just a tweaked form of Proof of Stake. Under this approach, only parties that have been approved are chosen based on their reputation. Proof of Authenticity is used by Ethereum's Koven Testnet Blockchain and IBM's Hyperledger Fabric. Two key characteristics of an effective consensus method are persistence and liveliness. The system's consistent response regarding the status of a transaction is ensured by persistence. When all nodes concur on a choice or a value, the system is said to be lively [48].

9.2.5 Blockchain limitations

Blockchain has shown itself to be a robust solution with a wide range of uses in the healthcare sector. Nevertheless, in order to fully utilize its potential, it is imperative to acknowledge and resolve its limitations. Table 9.2 lists the main obstacles to blockchain deployment in order to shed light on these restrictions and possible fixes.

9.2.6 Motivation of blockchain

In the realm of healthcare information systems, managing sensitive patient data and ensuring swift, secure access to healthcare information are paramount for delivering enhanced patient care. Leveraging blockchain technology as a secure distributed ledger offers several advantages for facilitating the exchange of medical data between systems.

First, patient charts are meticulously organized chronologically on the blockchain. This includes comprehensive records spanning outpatient, inpatient, and wearable device tests, all automatically sorted in chronological

Table 9.2 The limits of blockchain

Articles	*Subjects*	*Limitations*
[8]	Lack of standardization	The central limitation highlighted is the lack of standardization, which is identified as a significant obstacle to the widespread acceptance of blockchain technology.
[49]	Decentralized storage and privacy leakage	The primary limitation highlighted is the potential privacy leakage in decentralized systems, particularly as it pertains to healthcare settings and the stringent requirements of the sector. The need for users to enter private keys during the data retrieval process is identified as a potential source of privacy concerns.
[50]	Key management	The key limitation discussed is the challenge of devising effective key management principles for blockchain, addressing both security concerns and the practicality of key storage and recovery. The paper emphasizes the potential risks associated with compromised keys in the context of blockchain data security.
[51]	Scalability and IoT overhead	The key limitation highlighted is the scalability challenge within blockchain systems, particularly as it relates to the increasing computational demands in an IoT environment. The paper points out that this limitation may impede the seamless expansion of blockchain systems as the number of participants grows.
[52]	Scalability and Storage Capacity	The paper discusses two concerns related to storing information on the blockchain; privacy and the ability to handle an amount of data. Maintaining confidentiality is a challenge, due to everyone being able to see the data on the chain particularly when it comes to details. Moreover managing records such, as patient histories and test outcomes might overload the blockchain storage capabilities affecting its scalability adversely.
[53, 54]	Blockchain-specific vulnerabilities	The key limitation highlighted is the existence of specific vulnerabilities within blockchain technology, encompassing various attacks and potential threats. The paper underscores the intricate dynamics of the consensus mechanism, the computational requirements of miners, and the nuances of block withholding attacks.
[55]	Lack of universally-defined standards	The challenges of integrating technology into the healthcare industry arise from its developmental phase and the absence of well-defined standards. This gap calls for dedication and resources with a need for recognized standards to handle concerns, like data volume, structure, and formats compatible with blockchain storage. Setting up these benchmarks is vital, for facilitating the integration and implementation of blockchain within organizations.

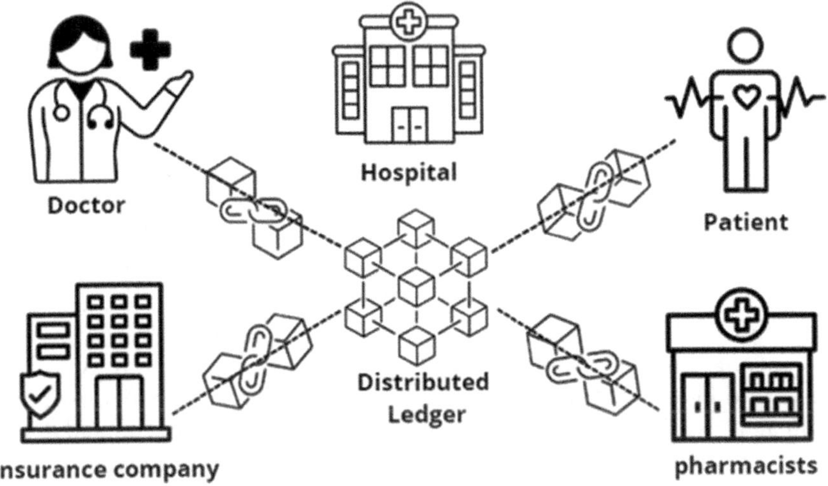

Figure 9.2 Benefit of using blockchain on the privacy of healthcare data.

order. Such an organization enables physicians and patients alike to review records seamlessly and obtain detailed treatment instructions.

Second, the blockchain enables decentralized data verification originating from a centralized source. In instances where the network opts for consensus, the blockchain facilitates public validation statements, thereby allowing the network to operate without the need for centralized authority overseeing its operations (Figure 9.2).

Ensuring guaranteed integrity and secure storage is another advantage of utilizing blockchain technology. A blockchain hash chain serves as a robust method for securely storing medical data. Even in the event of a few nodes being compromised by an adversary, the distributed nature of the healthcare blockchain ensures continued functionality [56].

Moreover, blockchain technology facilitates effective data sharing through interoperability. By providing scalable and reliable storage and access to medical data across institutional and administrative boundaries, blockchain enables seamless interoperability. This is achieved through the utilization of smart contracts and user-friendly application programming interfaces (APIs). With the adoption of blockchain in healthcare, the current challenges related to costs, delays, and administrative hurdles in medical data sharing and coordination among independent healthcare institutions can be significantly reduced, if not entirely eliminated. Consequently, blockchain offers a scalable and secure decentralized platform for obtaining, storing, and sharing medical records.

Since the inception of Bitcoin [45], blockchain technology has seen widespread adoption in digital finance, the financial sector, and corporate operations. Consequently, it has made its way into the healthcare sector as well.

Despite still being in its early stages, stakeholders within the healthcare ecosystem are actively developing and implementing blockchain technology [56].

We used a strict methodology to address our investigation into blockchain technology in the healthcare industry. Our methodology comprised multiple pivotal phases. Initially, in order to comprehend the new developments, uses, and difficulties associated with blockchain technology in the healthcare industry, we carried out a thorough analysis of academic and technical literature. Subsequently, we determined crucial analysis axes, such as confidentiality, security, and safe transactions, that directed our evaluation of blockchain's influence on patient-centered systems. We then carried out a thorough strategic analysis of blockchain adoption in healthcare using the SWOT framework. At the same time, we compared the features, advantages, and drawbacks of the several blockchain solutions that are available for healthcare institutions. Ultimately, we deliberated on the consequences of our discoveries and devised suggestions for legislators, scholars, and professionals operating in the healthcare field. We were able to offer a thorough and knowledgeable study of how blockchain technology might improve security, privacy, and safe transactions in patient-centered healthcare systems, thanks to our technique.

9.3 HEALTHCARE BLOCKCHAIN: OVERVIEW

The blockchain technology due to the decentralized principles it holds can ensure the confidential and authorized patient information that will give opportunities to the healthcare sector to have a shift. Accordingly, the present system of healthcare doctrine will collapse and a new system where people will have the responsibility of their own care arises. In the old order of a conventional healthcare system, medical records are vulnerable as they normally do not have reliable networks of Java Community Process (JCP), thereby allowing them to be readily hacked due to giant data breaches.

Another issue is that, despite the introduction of wearables and individualized treatment, people still do not have total ownership over their medical data. It is necessary to consider the significant moral ramifications of each of these circumstances.

Since the introduction of Industry 4.0 and other cutting-edge technology, the healthcare sector has grown tremendously. Electronic medical records (EMRs) now lead the way for patient data management and medical research, which were once handled through the paper approach. Two revolutionary healthcare tools are telemedicine and IoMT, invented lately. Blockchain is a tool that can lead to a radical overhaul of the medical world. The latest changes to healthcare systems are seen in Figure 9.3.

The application of blockchain technology is a manifold in the healthcare supply chain (HCSC), which comprises manufacturers of raw materials, logistics service providers, producers, wholesalers, retail distribution, and service providers. It also includes EHR, patient information validation, and data

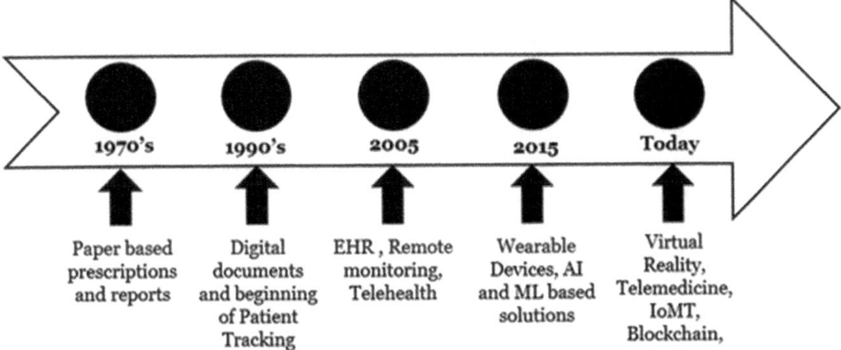

Figure 9.3 Evolution of healthcare systems.

analysis systems [57]. Healthcare applications of blockchain technology are numerous and varied and include things like improved mobile health apps, team monitoring, medical records sharing and storage, data archiving, clinical trials, and insurance data storage, among others [58]. Figure 9.4 illustrates the numerous applications of blockchain within the healthcare sector.

9.3.1 Drug traceability

A great deal of people suffer injuries or lose their lives each year as a result of using fake drugs. Counterfeiting pharmaceuticals results in serious financial and reputational consequences for pharmaceutical companies. By eliminating

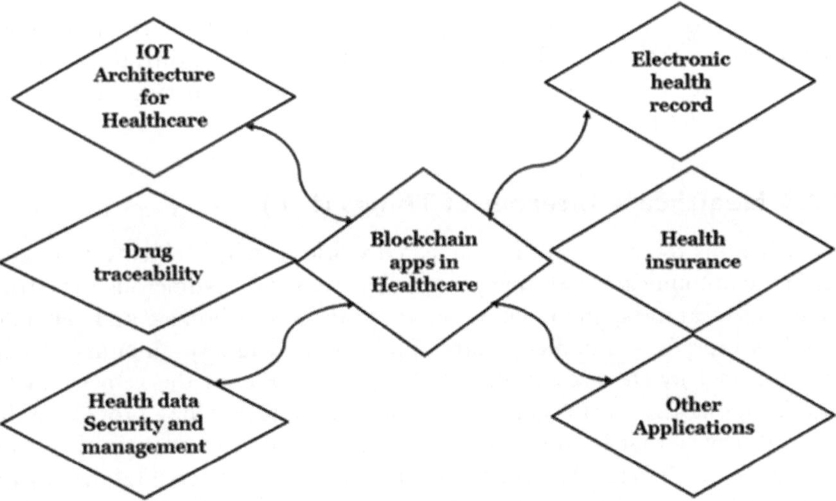

Figure 9.4 Applications of blockchain in healthcare system.

fraud and counterfeit goods, blockchain-based medicine traceability improves supply chain efficiency as opposed to using a traditional supply chain management system [34].

In [34] created a blockchain using G-coin to secure public health data and facilitate transparent medicine sales. In the realm of decentralized applications (dApps), there exists a front-end layer connected to a decentralized storage system, smart contracts, and on-chain resources through an API. Stakeholders are expected to access these resources using software devices. In [39] was developed a medicine traceability system, facilitating stakeholders' interactions with the medication monitoring system and smart contract [39].

9.3.2 Electronic medical records (EMR)

EMRs contain sensitive and private patient information and are typically stored within hospital systems. However, patients often encounter challenges in accessing their own health information, which may also be shared with multiple parties. To address this issue, The article [59] introduced Program for Advanced Computer Experiments (PACEX), a blockchain-based system that grants patients complete ownership over their EMRs. PACEX stores EMR hash values on the blockchain for integrity verification and tracks all EMR transfers while minimizing on-chain data storage. Implemented on the Ethereum private blockchain, PACEX comprises three main components: the blockchain itself, an application tailored for hospitals, and another application designed for patients [59].

The first component empowers patients to manage EMRs across multiple hospitals and offers them full control over their data access. The second component enables each hospital to process access requests and retrieve EMRs from other hospitals. To connect to the private blockchain network, each hospital operates an Ethereum node. Through the utilization of smart contracts, the blockchain records every interaction occurring between the involved parties [59].

9.3.3 Healthcare Internet of Things (IoT)

The IoT aims to create a fully connected world in which objects may interact and communicate with one another via data [60]. Numerous industries, including wearables, smart cities, smart grids, smart homes, and the automobile sector, have adopted smart applications. Massive amounts of data are produced by the growth of IoT. Keeping an eye on the veracity of the data is among the most crucial tasks. Blockchain technology provides a distributed service that is approved by all parties involved and guarantees that the data is unchangeable, which can be used to secure creditability in IoT data. Blockchain technology holds the promise of enhancing IoT data security and transparency, while also offering opportunities for improved efficiency, scalability, and standardization within the healthcare sector. In healthcare, the utilization of blockchain-enabled IoT devices empowers patients to regulate access to the data they generate, thereby safeguarding devices from cyber threats and monitoring information access. Moreover, blockchain-powered smart contract systems have the potential to automate supply chain payments based on real-time data detected by IoT sensors.

The integration of sensors into small devices for monitoring patient health status and data analysis has sparked considerable interest in leveraging IoT to support patients and healthcare professionals. This advancement may pave the way for the transformation of hospitals into "smart hospitals," streamlining various medical procedures for staff members. A secure data transfer method for blockchain-enabled IoT healthcare systems has been proposed in recent research [61], leveraging zero-knowledge evidence to ensure secure data transport and integrity. Additionally, Ethereum smart contracts offer solutions to data security concerns associated with interplanetary file systems, while strategies such as blockchain networking, as suggested by other studies [62], prioritize privacy and confidentiality for healthcare data shared over the blockchain. The approaches dealing with these problems also tackle the problems that most healthcare blockchains have, which are identification, consensus, throughput, latency, and security levels.

In addition, the application of smart particles and multimedia in healthcare structure for the sake of analysis, storage, and online sharing of patient data in text, voice, and image formats is also one of the important feature. To tackle issues related to data security and management in healthcare, recent research [63] proposes an IoT multimedia data processing system specifically tailored for the healthcare sector. This system aims to address concerns surrounding diagnosis enrollment and ensure the safe and efficient processing of transactions.

In another innovative approach, private blockchain, coupled with federated learning, is recommended by researchers [64] to address privacy and security challenges associated with fog-IoT. This fog-IoT network focuses on overcoming challenges related to wearable IoT devices, including adaptability, privacy, integrity, and security.

9.3.4 Patient data management (EHR)

Lab tests are performed, patient visits are organized, billing and accounts are managed, and medical data is electronically maintained. They mainly aim to provide a more accessible and secure storage platform for patient data considering the aspect of cost-effectiveness [65]. The system ties up with the EHRs which are used in the health care industry all over. The aim here is to ensure the transmission of the medical record in a secure and immutable manner across the digital space. By preserving each patient's complete medical history on the blockchain, this system facilitates multi-level management by patients, physicians, regulators, hospitals, insurers, and other stakeholders.

This approach provides a secure method for documenting and storing the comprehensive medical histories of all patients.

9.3.5 Health insurance: Claims processing

Health insurance has become essential in people's lives due to the increase in health difficulties. Emergency medical care may be challenging for people with low incomes. Health insurance provides financial protection from debt worry and aids in covering medical expenses in case of an emergency. Concerns about fraud, privacy, and security may surface in relation to health insurance and all of its benefits. Since fraud may result in huge losses for people, companies, and governments, it has become a sensitive problem in the health insurance market in recent years. For this reason, national authorities as well as commercial firms must create systems that can recognize fraudulent instances and payments. of the most annoying procedures is filing an insurance claim, as insurance processors must go through a vast amount of data to look for false claims, segmented data sources, and canceled policies that the customers report. With the use of this technology, risk can be managed transparently, and insurers can now acquire ownership of the assets they are insuring [66]. These days, a lot of the insurance system depends on transactions. Blockchain technology has produced an immutable record that many Fintech systems utilize to detect cryptocurrency double-spending in order to meet efficiency and security criteria [66]. A blockchain-based insurance system offers two distinct insurance service models: there is an assurance aspect for the protection of the web identity security of users or, again, to be implemented particularly to make it more data secure for commercial websites. The claim will be submitted and followed at once with the verification of the documents to substantiate the matter brought forward. Contracts that are being enforced into play it is only automatically will not make any hassle for the policyholder as well as the insurer [67].

9.3.6 Supply chain management

Healthcare providers need to efficiently handle the HCSC procedure in their daily operations as well as during pandemics such as COVID-19. Despite significant advancements in new technologies and treatment alternatives, healthcare practitioners continue to grapple with challenges in purchasing, ordering, forecasting, and distributing substantial volumes of cargo on a daily basis, both domestically and globally. The healthcare sector, akin to manufacturing supply chains, faces significant obstacles in optimizing and managing its operations [58].

Group purchasing organizations (GPOs) become an important participants in the HCSC by helping the providers in vendor selection and providing multiple volume discounts. To enhance the efficiency of the HCSC process of contracting and to reduce the costs of administration of GPO, contracts are suggested through proposing a blockchain-based solution. This solution comprises two main components: an Ethereum smart contract and a decentralized storage system. The pharmaceutical industry is connected via a single decentralized Ethereum network that consists of all the links, distributors, and health providers throughout the entire purchase of medications and devices network. Trust and transparency are provided by the proposed architecture for smart contract access to be limited to registered parties only. Such a mechanism prevents price differences between stakeholders, thus ensuring lower cost.

An equitable and effective global vaccination campaign will only happen if data covering the networks has been audited by all the interested players. Different kinds of items in a vaccination will be prone to damage during transportation due to thermal instability, improper air and daylight control, and poor storage conditions. IoT is a field of application that has enabled end-to-end visibility and transparency in existing transportation channels. To recall, in the article [68], CryptoCargo was detailed as an intelligent container system. It is a blockchain-enabled platform that is mainly used to monitor shipping conditions and detect any transgressions that can endanger the goods transported.

9.3.7 Device tracking

Monitoring medical equipment for the entire lifecycle of that gadget right from its creation until it is decommissioned is the most important factor in uprooting healthcare. Hospitals and their employees, who are responsible for fleets of medical devices used in different units and by patients, experience difficulties finding the necessary materials quickly when dealing with sudden emergency situations. Subsequently, the traceability and supervision of medical units are crucial for fast device return, performance of device purchases, and battling the fraud in the supply chain, thus all the way to the phase of device discarding [69].

Comparing a blockchain-based system to conventional location monitoring methods reveals several advantages, with the most notable being blockchain's immutability and tamper-proof nature. Blockchain technology can create an immutable record that showcases crucial device information, including the producer, reseller, serial number, location, and history, aiding in regulatory compliance. This approach prevents malicious entities from altering or erasing a device's location history from the database.

By leveraging blockchain technology, we can ensure that clients utilize the allocated medical equipment while gaining valuable insights into its utilization and whereabouts. Benefits include streamlining the usage process, maintaining a comprehensive inventory process log, and facilitating easy access to equipment during emergencies.

9.4 SWOT ANALYSIS OF THE ADOPTION OF THE BLOCKCHAIN IN HEALTHCARE

Blockchain, when applied to the healthcare sector, brings forth significant advantages while also posing challenges. Its strengths, including disintermediation, automation, immutability, transparency, privacy, auditability, and efficient data management, have the potential to transform the management of medical information. By fostering trust through immutable and transparent transactions, blockchain can encourage enhanced cooperation among healthcare system operators, facilitating better care coordination and more efficient clinical data management.

However, these advantages do not come without their share of challenges. Operational costs, the possibility of creating forks, lack of flexibility, increased needs for local server storage capacity, scalability issues, and high energy consumption are weaknesses that need careful consideration. These aspects could pose financial and technical barriers to the widespread adoption of blockchain in the healthcare domain (Table 9.3).

On the opportunity front, improved collaboration among healthcare system operators, the rise in technical awareness, and the acquisition of new skills open the door to significant advancements. The potential for global health data exchange, coupled with telemedicine and remote patient monitoring, presents promising prospects for more efficient and accessible healthcare delivery, especially in remote regions.

However, the landscape is not without its threats. Resistance to change, lack of expertise and trust from healthcare professionals, along with the absence of standardization and vulnerability to cyberattacks, pose significant

Table 9.3 SWOT matrix

Positive	*Negative*
Strengths	*Weaknesses*
• Disintermediation and automation • Immutability • Trust • Transparency • Privacy • Auditability • Efficient Data Management	• Operation Costs • creation of a possible fork • Lack of flexibility • Need for greater capability of data storage on local servers • Scalability Issues • Energy Consumption
Opportunities	*Threats*
• Enhanced cooperation between healthcare system operators • Rise in technical awareness and acquisition of new skills • Global Health Data Exchange • Telemedicine and Remote Patient Monitoring	• Resistance to change • Lack of expertise • Lack of trust in the application of a new technology by healthcare workers. • Lack of Standardization • Vulnerability to Cyberattacks

challenges. These factors could impede the swift adoption of blockchain in healthcare, highlighting the need for increased awareness, training programs, and robust security protocols to ensure the long-term success of this innovative technology in the medical field.

9.5 COMPARATIVE STUDY: SUMMARIZING THE CHARACTERISTICS OF VARIOUS BLOCKCHAIN SOLUTIONS IN THE HEALTHCARE SECTOR

Table 9.4 provides insights into the application of blockchain technology in the healthcare sector, showcasing various solutions along with their notable characteristics, consensus algorithms, standards compliance, and healthcare use cases. Let's delve into several blockchain-based healthcare solutions for further examination.

9.6 DISCUSSION

Investigating blockchain technology in the healthcare industry opens up a world of possibilities, each with specific considerations and repercussions. Our analysis, which is shown in Table 9.4, demonstrates the wide range of blockchain solutions that are out there, each of which adds to a more complex comprehension of its possible uses and implications. Several consensus algorithms are included, ranging from the conventional PoW to the novel Proof of Elapsed Time (PoET), which calls for a critical assessment of their effectiveness, security, and energy usage in healthcare environments. Following laws like the General Data Protection Regulation (GDPR), the Health Insurance Portability and Accountability Act (HIPAA), and the National Institute of Standards and Technology (NIST) requirements is crucial. These rules emphasize how crucial it is to protect patient information and make sure that laws are followed. Still, the lack of full compliance standards in some solutions raises questions about how thorough regulatory adherence is, and more research is required.

Furthermore, the division between public and private blockchains and the incorporation of cryptocurrencies provide complex aspects of control, transparency, and accessibility. Private blockchains give better control and privacy, while public blockchains enable unmatched accessibility and openness. Though it raises concerns about risk management and financial governance, the incorporation of cryptocurrencies offers novel ways to incentivize and facilitate transactions.

Advancements in technology, such as gene token-based exchanges, hierarchical deterministic wallets, and smart contracts for data encryption, support security and privacy protocols in blockchain-enabled healthcare systems. These developments strengthen the infrastructure that protects private

Table 9.4 Characteristics of various blockchain solutions in the healthcare sector

Art	*Blockchain solution*	*Key features*	*Consensus algorithm*	*Compliance*	*Use case*
[70]	Blockchain Health	On-chain and off-chain data storage, identity verification through key pairs, PoET consensus algorithm, HIPAA compliant.	Proof of Elapsed Time (PoET)	HIPAA	Data sharing between patients and researchers.
[71]	Burst IQ	Blockchain-based healthcare data-driven platforms, health wallet for individual life graphs, interactive big data distribution, machine learning integration, HIPAA, GDPR, NIST compliant.	Not specified	HIPAA, GDPR, NIST	Individual data sharing, management, sale, and donation.
[72]	MedRec	Patient data authentication and sharing, Proof of Work (PoW) mining, incentives for data sharing, transparent audibility, decentralized model, HIPAA compliant.	Proof of Work (PoW)	HIPAA	Patient data authentication and sharing platform.
[73]	Gem Health	Revenue management through data permission and sharing, real-time and transparent health claim system, and collaboration with Philips.	Not specified	Not specified	Real-time health claim processing and data transfer.
[74]	Model chain	Model chain architecture for cross-institutional collaboration, private blockchain for metadata management, and integration of machine learning.	Not specified	Not specified	Facilitation of Patient-Centered Outcomes Research (PCOR) and interinstitutional collaboration.
[75]	Guard time	Transparent patient record management, collaboration with Estonian eHealth, Oracle, and blockchain database usage.	Not specified	Not specified	Transparency and audibility of patient records.

(*Continued*)

Table 9.4 (*Continued*)

Art	*Blockchain solution*	*Key features*	*Consensus algorithm*	*Compliance*	*Use case*
[76]	HealthCombix	Decentralized healthcare ecosystem, token-based privacy-preserving patient data management, disease prediction, and transparent data asset monetization.	Private blockchain	Not specified	Peer-to-peer communication, disease prediction, data asset monetization.
[77]	IBM Blockchain	Automatic clinical trial, transparent health record sharing, IBM's healthcare blockchain applications.	Not specified	Not specified	Automatic clinical trial, transparent health record sharing.
[78]	Universal health Coin	Blockchain and AI-based cryptocurrency, encrypted and secured transactions, data exchange among stakeholders using UHC coin.	Public-Private Blockchain Key	Not specified	Exchange of data among stakeholders using UHC coin
[79]	Genomes	Off-chain and on-chain gene information storing, GENE tokens as a medium of exchange, securely sharing biological information.	Not specified	Not specified	Secure sharing of gene and biological information.
[80]	Youbase	Hierarchical deterministic (HD) based wallet, tree-like structure with keys, data anonymization technique, community-based network architecture.	Not specified	Not specified	Controlled access to personal information with data anonymization.
[81]	HapiChain	Comprehensive telemedicine framework with modularity, security, reliability, and IPFS-based scalable medical file storage. Content-based access controls that can be customized for medical records.	Public Blockchain	Not specified	Teleconsultation and telemonitoring integration for comprehensive remote treatment.

(*Continued*)

Table 9.4 (Continued)

Art	*Blockchain solution*	*Key features*	*Consensus algorithm*	*Compliance*	*Use case*
[82]	Not specified	Immutable ledger for data integrity and transparency, Decentralized system for enhanced security and privacy, smart contracts for automated and secure transactions	Public blockchain	Not specified	Implementation in healthcare institutions to improve EHR management and data security
[83]	Not specified	Interleaving encoder for EMR encryption (t, n)-threshold message sharing scheme for data reconstruction, Smart contract integration for secure processes	Proof of Work	Not specified	Cross-institutional sharing of EMRs, efficient data storage, and retrieval in healthcare blockchain

patient data and guarantees data integrity throughout the healthcare system. Furthermore, the incorporation of blockchain technology with telemedicine, as demonstrated by initiatives such as HapiChain [81], offers revolutionary prospects for the provision of remote healthcare services. However, the adoption of blockchain-enabled telemedicine systems is hampered by issues with scalability, security, and ethics. To address these issues and guarantee that everyone has fair access to high-quality healthcare services, strong frameworks that strike a balance between innovation and accountability must be developed. When examining the potential future direction of blockchain technology in the healthcare industry, it is critical to recognize the inherent constraints and difficulties. Although our analysis offers a thorough summary of recent breakthroughs, it also emphasizes the necessity of further research and development. To fully utilize blockchain in healthcare, future research should focus on issues like interoperability, scalability, and governance frameworks. To sum up, the field of blockchain solutions in healthcare offers a diverse range of prospects and obstacles that influence the advancement of innovation within the sector. We may develop a comprehensive knowledge of blockchain's consequences in healthcare by adopting a multifaceted approach that takes into account the opinions of patients, healthcare providers, insurers, and other stakeholders. By carefully examining potential future directions and openly discussing existing constraints, we can steer toward a blockchain-powered healthcare ecosystem that is more transparent, safe, and patient-focused.

9.7 CONCLUSION

The integration of blockchain technology presents a promising avenue for addressing persistent challenges within the healthcare industry. By leveraging blockchain's foundational principles of security, integrity, decentralization, and authentication, healthcare stakeholders can explore novel solutions to enhance data management and exchange. However, the rapid expansion of technical infrastructure, including Internet-enabled devices and IoT technologies, has introduced new vulnerabilities that malicious actors may exploit to compromise data integrity and patient privacy. This underscores the urgency for robust security measures and innovative approaches to safeguard sensitive healthcare data.

Research indicates that blockchain holds the potential to mitigate existing challenges within the healthcare landscape. Empowering patients with ownership and control over their medical data, blockchain offers a secure platform for data sharing while ensuring privacy and confidentiality. Through a comprehensive analysis of blockchain's architecture, applications, consensus techniques, and security considerations, this study sheds light on the transformative potential of blockchain in healthcare.

Nevertheless, it is essential to recognize that blockchain is not a panacea for all healthcare-related issues. Rather, a nuanced understanding of specific challenges and their implications for the industry is imperative. Critical aspects such as mining incentives and potential blockchain vulnerabilities warrant further investigation to fully comprehend blockchain's impact on healthcare systems.

Furthermore, while blockchain and AI integration holds promise for advancing patient care and accessibility to healthcare services, challenges such as budget constraints and a shortage of expertise pose significant barriers to widespread adoption. Addressing these obstacles requires concerted efforts to enhance awareness, build expertise, and foster collaboration within the healthcare ecosystem.

In conclusion, while blockchain technology offers exciting opportunities to revolutionize healthcare, its implementation must be approached thoughtfully and collaboratively. By leveraging blockchain's strengths and addressing its limitations, healthcare organizations can navigate the complexities of data management and drive innovation in patient care delivery.

REFERENCES

[1] Narayanan, A., Bonneau, J., Felten, E., Miller, A., and Goldfeder, S., *Bitcoin and Cryptocurrency Technologies: A Comprehensive Introduction*, Princeton, NJ: Princeton University Press, 2016.

[2] Yue, X., Wang, H., Jin, D., Li, M., and Jiang, W., 'Healthcare Data Gateways: Found Healthcare Intelligence on Blockchain with Novel Privacy Risk Control', *Journal of Medical Systems*, vol. 40, no. 10, p. 218, Oct. 2016. doi: 10.1007/s10916-016-0574-6.

[3] Zhang, P., Schmidt, D. C., White, J., and Lenz, G., 'Blockchain Technology Use Cases in Healthcare', in *Advances in Computers*, vol. 111, Elsevier, 2018, pp. 1–41. Accessed: Mar. 11, 2024. [Online]. Available: https://www.sciencedirect.com/science/article/pii/S0065245818300196.

[4] Treleaven, P., Brown, R. G., and Yang, D., 'Blockchain Technology in Finance', *Computer*, vol. 50, no. 9, pp. 14–17, 2017.

[5] Liu, C., Xiao, Y., Javangula, V., Hu, Q., Wang, S., and Cheng, X., 'NormaChain: A Blockchain-Based Normalized Autonomous Transaction Settlement System for IoT-Based E-commerce', *IEEE Internet of Things Journal*, vol. 6, no. 3, pp. 4680–4693, 2018.

[6] Miller, D., 'Blockchain and the Internet of Things in the Industrial sector', *IT Professional*, vol. 20, no. 3, pp. 15–18, 2018.

[7] Fu, Y., and Zhu, J., 'Big Production Enterprise Supply Chain Endogenous Risk Management Based on Blockchain', *IEEE Access*, vol. 7, pp. 15310–15319, 2019.

[8] Ølnes, S., Ubacht, J., and Janssen, M., 'Blockchain in Government: Benefits and Implications of Distributed Ledger Technology for Information Sharing', *Government Information Quarterly*, vol. 34, no. 3., pp. 355–364, 2017. Accessed: Mar. 11, 2024. [Online]. Available: https://www.sciencedirect.com/science/article/pii/S0740624X17303155.

[9] Carson, B., Romanelli, G., Walsh, P., and Zhumaev, A., 'Blockchain beyond the Hype: What Is the Strategic Business Value', *McKinsey & Company*, vol. 1, pp. 1–13, 2018.

[10] Ramachandran, A., and Kantarcioglu, D. M., 'Using Blockchain and smart contracts for secure data provenance management', *arXiv*, Sep. 28, 2017. Accessed: Mar. 11, 2024. [Online]. Available: http://arxiv.org/abs/1709.10000.

[11] Zhu, L., Wu, Y., Gai, K., and Choo, K.-K. R., 'Controllable and Trustworthy Blockchain-Based Cloud Data Management', *Future Generation Computer Systems*, vol. 91, pp. 527–535, 2019.
[12] Treiblmaier, H., and Sillaber, C., 'The Impact of Blockchain on E-Commerce: A Framework for Salient Research Topics', *Electronic Commerce Research and Applications*, vol. 48, p. 101054, 2021.
[13] Hollander, M. J., Chappell, N. L., Prince, M. J., and Shapiro, E., 'Providing Care and Support for an Aging Population: Briefing Notes on Key Policy issues', *Aging Clinical and Experimental Research*, vol. 15, no. 3 Suppl, pp. 1–2, 2007.
[14] Halim, A. H. A., Halim, M. H. A., and Usman, S., 'Implementation of IoT and Blockchain for Temperature Monitoring in Covid-19 Vaccine Cold Chain Logistics', *Open International Journal of Informatics*, vol. 9, no. 1, pp. 78–87, 2021.
[15] Mahmood, Z., 'Impact of Blockchain technology in healthcare sector during COVID-19 pandemic', in *Computer Science & Information Technology: International Conference on AI, Machine Learning and Applications (AIMLA 2021)*, August 28–29, 2021, Dubai, UAE, AIRCCpublishing Corporation, 2021, pp. 75–88. Accessed: Mar. 11, 2024. [Online]. Available: https://epubl.ktu.edu/object/elaba:105135511/.
[16] Khatri, S., Alzahrani, F. A., Ansari, M. T. J., Agrawal, A., Kumar, R., and Khan, R. A., 'A Systematic Analysis on Blockchain Integration with Healthcare Domain: Scope and Challenges', *IEEE Access*, vol. 9, pp. 84666–84687, 2021.
[17] Islam, A., and Shin, S. Y., 'A Blockchain-Based Secure Healthcare Scheme with the Assistance of Unmanned Aerial Vehicle in Internet of Things', *Computers & Electrical Engineering*, vol. 84, p. 106627, 2020.
[18] Koteska, B., Karafiloski, E., and Mishev, A., 'Blockchain implementation quality challenges: A literature', in *SQAMIA 2017: 6th Workshop of Software Quality, Analysis, Monitoring, Improvement, and Applications*, 2017, p. 2017. Accessed: Mar. 11, 2024. [Online]. Available: https://www.academia.edu/download/57473742/paper-kot.pdf
[19] Gökalp, E., Gökalp, M. O., Çoban, S., and Eren, P. E., 'Analysing Opportunities and Challenges of Integrated Blockchain Technologies in Healthcare', in *Information Systems: Research, Development, Applications, Education*, vol. 333, S. Wrycza and J. Maślankowski, Eds., in Lecture Notes in Business Information Processing, vol. 333., Cham: Springer International Publishing, 2018, pp. 174–183. doi: 10.1007/978-3-030-00060-8_13.
[20] Agbo, C. C., Mahmoud, Q. H., and Eklund, J. M., 'Blockchain technology in healthcare: A systematic review', in *Healthcare*, MDPI, 2019, p. 56. Accessed: Mar. 11, 2024. [Online]. Available: https://www.mdpi.com/2227-9032/7/2/56.
[21] Puthal, D., Malik, N., Mohanty, S. P., Kougianos, E., and Das, G., 'Everything You Wanted to Know about the Blockchain: Its Promise, Components, Processes, and Problems', *IEEE Consumer Electronics Magazine*, vol. 7, no. 4, pp. 6–14, 2018.
[22] Kuo, T.-T., Kim, H.-E., and Ohno-Machado, L., 'Blockchain Distributed Ledger Technologies for Biomedical and Health Care Applications', *Journal of the American Medical Informatics Association*, vol. 24, no. 6, pp. 1211–1220, 2017.
[23] Sullivan, C., and Burger, E., 'E-Residency and Blockchain', *Computer Law & Security Review*, vol. 33, no. 4, pp. 470–481, 2017.
[24] Anjum, A., Sporny, M., and Sill, A., 'Blockchain Standards for Compliance and Trust', *IEEE Cloud Computing*, vol. 4, no. 4, pp. 84–90, 2017.
[25] Cai, Y., and Zhu, D., 'Fraud Detections for Online Businesses: A Perspective from Blockchain Technology', *Financial Innovation*, vol. 2, no. 1, p. 20, 2016. doi: 10.1186/s40854-016-0039-4.

[26] Zhu, H., and Zhou, Z. Z., 'Analysis and Outlook of Applications of Blockchain Technology to Equity Crowdfunding in China', *Financial Innovation*, vol. 2, pp. 1–11, 2016.
[27] Shrestha, A. K., and Vassileva, J., 'Towards decentralized data storage in general cloud platform for meta-products', in *Proceedings of the International Conference on Big Data and Advanced Wireless Technologies*, Blagoevgrad Bulgaria: ACM, Nov. 2016, pp. 1–7. doi: 10.1145/3010089.3016029.
[28] Karafiloski, E., and Mishev, A., 'Blockchain solutions for big data challenges: A literature review', in *IEEE EUROCON 2017-17th International Conference on Smart Technologies*, IEEE, 2017, pp. 763–768. Accessed: Mar. 11, 2024. [Online]. Available: https://ieeexplore.ieee.org/abstract/document/8011213/.
[29] Stanciu, A., 'Blockchain based distributed control system for edge computing', in *2017 21st International Conference on Control Systems and Computer Science (CSCS)*, IEEE, 2017, pp. 667–671. Accessed: Mar. 11, 2024. [Online]. Available: https://ieeexplore.ieee.org/abstract/document/7968630/.
[30] Xia, Q. I., Sifah, E. B., Asamoah, K. O., Gao, J., Du, X., and Guizani, M., 'MeDShare: Trust-Less Medical Data Sharing among Cloud Service Providers via Blockchain', *IEEE Access*, vol. 5, pp. 14757–14767, 2017.
[31] Suzuki, S., and Murai, J., 'Blockchain as an audit-able communication channel', in *2017 IEEE 41st Annual Computer Software and Applications Conference (COMPSAC)*, Vol. 2, IEEE, pp. 516–522, 2017.
[32] Underwood, S., 'Blockchain beyond Bitcoin', *Communications of the ACM*, vol. 59, no. 11, pp. 15–17, Oct. 2016. doi: 10.1145/2994581.
[33] Fontana, P., Diirr, B., and et Cappelli, C., 'Transparency Challenges in Blockchain', EGOV-CeDEM-ePart, p. 193, 2018.
[34] Isravel, D. P., Sagayam, K. M., Bhushan, B., Sei, Y., and Eunice, J., 'Blockchain for Healthcare Systems: Architecture, Security Challenges, Trends and Future Directions', *Journal of Network and Computer Applications*, vol. 215, p. 103633, Jun. 2023. doi: 10.1016/j.jnca.2023.103633.
[35] Gordon, W. J., and Catalini, C., 'Blockchain Technology for Healthcare: Facilitating the Transition to Patient-Driven Interoperability', *Computational and Structural Biotechnology Journal*, vol. 16, pp. 224–230, Jan. 2018. doi: 10.1016/j.csbj.2018.06.003.
[36] Fanning, K., and Centers, D. P., 'Blockchain and Its Coming Impact on Financial Services', *Journal of Corporate Accounting & Finance*, vol. 27, no. 5, pp. 53–57, Jul. 2016. doi: 10.1002/jcaf.22179.
[37] Tschorsch, F., and Scheuermann, B., 'Bitcoin and Beyond: A Technical Survey on Decentralized Digital Currencies', *IEEE Communications Surveys & Tutorials*, vol. 18, no. 3, pp. 2084–2123, 2016. doi: 10.1109/COMST.2016.2535718.
[38] Benchoufi, M., Porcher, R., and Ravaud, P., 'Blockchain Protocols in Clinical Trials: Transparency and Traceability of Consent', *F1000Res*, vol. 6, p. 66, Feb. 2018. doi: 10.12688/f1000research.10531.5.
[39] Musamih, A., Jayaraman, R., Salah, K., Hasan, H. R., Yaqoob, I., and Al-Hammadi, Y., 'Blockchain-Based Solution for Distribution and Delivery of COVID-19 Vaccines', *IEEE Access*, vol. 9, pp. 71372–71387, 2021. doi: 10.1109/ACCESS.2021.3079197.
[40] Pop, C., Cioara, T., Antal, M., Anghel, I., Salomie, I., and Bertoncini, M., 'Blockchain Based Decentralized Management of Demand Response Programs in Smart Energy Grids', *Sensors*, vol. 18, no. 1, Art. no. 1, Jan. 2018. doi: 10.3390/s18010162.
[41] Al-Karaki, J. N., Gawanmeh, A., Ayache, M., and Mashaleh, A., 'DASS-CARE: A decentralized, accessible, scalable, and secure healthcare framework using blockchain', in *2019 15th International Wireless Communications & Mobile Computing Conference (IWCMC)*, Jun. 2019, pp. 330–335. doi: 10.1109/IWCMC.2019.8766714.

[42] Liu, X., Wang, Z., Jin, C., Li, F., and Li, G., 'A Blockchain-Based Medical Data Sharing and Protection Scheme', *IEEE Access*, vol. 7, pp. 118943–118953, 2019. doi: 10.1109/ACCESS.2019.2937685.

[43] Chelladurai, U., Pandian, S., and Ramasamy, K., 'A Blockchain Based Patient Centric Electronic Health Record Storage and Integrity Management for E-Health Systems', *Health Policy and Technology*, vol. 10, no. 4, p. 100513, Dec. 2021. doi: 10.1016/j.hlpt.2021.100513.

[44] Zheng, Z., Xie, S., Dai, H., Chen, X., and chen, X., 'An Overview of Blockchain Technology: Architecture, Consensus, and Future Trends', IEEE Conference Publication, IEEE Xplore. Accessed: Mar. 11, 2024. [Online]. Available: https://ieeexplore.ieee.org/abstract/document/8029379.

[45] Nakamoto, S., 'Bitcoin: A peer-to-peer electronic cash system', 2008.

[46] Androulaki, E. *et al.*, 'Hyperledger fabric: A distributed operating system for permissioned blockchains', in *Proceedings of the Thirteenth EuroSys Conference*, in EuroSys '18. New York, NY, USA: Association for Computing Machinery, Apr. 2018, pp. 1–15. doi: 10.1145/3190508.3190538.

[47] Ripple, 'Ripple—One frictionless experience to send money globally', 2018. Accessed: Sept. 4, 2018. [Online]. Available: https://ripple.com/.

[48] Swarnkar, M., Bhadoria, R. S., and Sharma, N., 'Security, Privacy, Trust Management and Performance Optimization of Blockchain Technology', in *Applications of Blockchain in Healthcare*, S. Namasudra and G. C. Deka, Eds., in Studies in Big Data., Singapore: Springer, 2021, pp. 69–92. doi: 10.1007/978-981-15-9547-9_3.

[49] Liang, X., Shetty, S., Tosh, D., Kamhoua, C., Kwiat, K., and Njilla, L., 'ProvChain: A blockchain-based data provenance architecture in cloud environment with enhanced privacy and availability', in *2017 17th IEEE/ACM International Symposium on Cluster, Cloud and Grid Computing (CCGRID)*, May 2017, pp. 468–477. doi: 10.1109/CCGRID.2017.8.

[50] Zhao, H., Zhang, Y., Peng, Y., and Xu, R., 'Lightweight backup and efficient recovery scheme for health blockchain keys', in *2017 IEEE 13th International Symposium on Autonomous Decentralized System (ISADS)*, Mar. 2017, pp. 229–234. doi: 10.1109/ISADS.2017.22.

[51] 'OmniPHR: A distributed architecture model to integrate personal health records', *Science Direct*. Accessed: Mar. 11, 2024. [Online]. Available: https://www.sciencedirect.com/science/article/pii/S1532046417301089.

[52] Pirtle, C., and Ehrenfeld, J., 'Blockchain for Healthcare: The Next Generation of Medical Records?', *Journal of Medical Systems*, vol. 42, no. 9, p. 172, Aug. 2018. doi: 10.1007/s10916-018-1025-3.

[53] Tosh, D. K., Shetty, S., Liang, X., Kamhoua, C. A., Kwiat, K. A., and Njilla, L., 'Security implications of blockchain cloud with analysis of block withholding attack', in *2017 17th IEEE/ACM International Symposium on Cluster, Cloud and Grid Computing (CCGRID)*, May 2017, pp. 458–467. doi: 10.1109/CCGRID.2017.111.

[54] 'Are blockchains immune to all malicious attacks?' *Financial Innovation*. Accessed: Mar. 11, 2024. [Online]. Available: https://link.springer.com/article/10.1186/s40854-016-0046-5.

[55] Siyal, A. A., Junejo, A. Z., Zawish, M., Ahmed, K., Khalil, A., and Soursou, G., 'Applications of Blockchain Technology in Medicine and Healthcare: Challenges and Future Perspectives', *Cryptography*, vol. 3, no. 1, Art. no. 1, Mar. 2019. doi: 10.3390/cryptography3010003.

[56] 'Security and privacy for healthcare blockchains', *IEEE Journals & Magazine | IEEE Xplore*. Accessed: Mar. 11, 2024. [Online]. Available: https://ieeexplore.ieee.org/abstract/document/9445631.

[57] Singh, A. P. *et al.*, 'A Novel Patient-Centric Architectural Framework for Blockchain-Enabled Healthcare Applications', *IEEE Transactions on Industrial Informatics*, vol. 17, no. 8, pp. 5779–5789, Aug. 2021. doi: 10.1109/TII.2020.3037889.

[58] 'Automating procurement contracts in the healthcare supply chain using blockchain smart contracts', *IEEE Journals & Magazine | IEEE Xplore*. Accessed: Mar. 11, 2024. [Online]. Available: https://ieeexplore.ieee.org/abstract/document/9363880.
[59] Ciampi, M., Esposito, A., Marangio, F., Sicuranza, M., and Schmid, G., 'Modernizing Healthcare by Using Blockchain', in *Applications of Blockchain in Healthcare*, S. Namasudra and G. C. Deka, Eds., in Studies in Big Data., Singapore: Springer, 2021, pp. 29–67. doi: 10.1007/978-981-15-9547-9_2.
[60] Reyna, A., Martín, C., Chen, J., Soler, E., and Díaz, M., 'On Blockchain and Its Integration With IoT. Challenges and Opportunities', *Future Generation Computer Systems*, vol. 88, pp. 173–190, Nov. 2018. doi: 10.1016/j.future.2018.05.046.
[61] 'A blockchain-orchestrated deep learning approach for secure data transmission in IoT-enabled healthcare system', *Science Direct*. Accessed: Mar. 11, 2024. [Online]. Available: https://www.sciencedirect.com/science/article/pii/S0743731522002106.
[62] Zhao, Z., Li, X., Luan, B., Jiang, W., Gao, W., and Neelakandan, S., 'Secure Internet of Things (IoT) Using a Novel Brooks Iyengar Quantum Byzantine Agreement-Centered Blockchain Networking (BIQBA-BCN) Model in Smart Healthcare', *Information Sciences*, vol. 629, pp. 440–455, Jun. 2023. doi: 10.1016/j.ins.2023.01.020.
[63] Taloba, A. I. *et al.*, 'A Blockchain-Based Hybrid Platform for Multimedia Data Processing in IoT-Healthcare', *Alexandria Engineering Journal*, vol. 65, pp. 263–274, Feb. 2023. doi: 10.1016/j.aej.2022.09.031.
[64] Baucas, M. J., Spachos, P., and Plataniotis, K. N., 'Federated Learning and Blockchain-Enabled Fog-IoT Platform for Wearables in Predictive Healthcare', *IEEE Transactions on Computational Social Systems*, vol. 10, no. 4, pp. 1732–1741, Aug. 2023. doi: 10.1109/TCSS.2023.3235950.
[65] Shahnaz, A., Qamar, U., and Khalid, A., 'Using blockchain for electronic health records', *IEEE Journals & Magazine | IEEE Xplore*. Accessed: Mar. 11, 2024. [Online]. Available: https://ieeexplore.ieee.org/abstract/document/8863359.
[66] Raikwar, M., Mazumdar, S., Ruj, S., Sen Gupta, S., Chattopadhyay, A., and Lam, K.-Y., 'A blockchain framework for insurance processes', in *2018 9th IFIP International Conference on New Technologies, Mobility and Security (NTMS)*, Feb. 2018, pp. 1–4. doi: 10.1109/NTMS.2018.8328731.
[67] Guo, Y. *et al.*, 'WISChain: An online insurance system based on blockchain and denglu1 for web identity security', in *2018 1st IEEE International Conference on Hot Information-Centric Networking (HotICN)*, Aug. 2018, pp. 242–243. doi: 10.1109/HOTICN.2018.8606011.
[68] 'Design and implementation of CryptoCargo: A blockchain-powered smart shipping container for vaccine distribution', *IEEE Journals & Magazine | IEEE Xplore*. Accessed: Mar. 11, 2024. [Online]. Available: https://ieeexplore.ieee.org/abstract/document/9395097.
[69] Jafri, R., and Singh, S., '4 – Blockchain Applications for the Healthcare Sector: Uses beyond Bitcoin', in *Blockchain Applications for Healthcare Informatics*, S. Tanwar, Ed., Academic Press, 2022, pp. 71–92. doi: 10.1016/B978-0-323-90615-9.00022-0.
[70] Smith, B., 'DokChain', 2016. Retrieved from: https://www.healthcareitnews.com/news/pokitdok-gains-healthcare-traction-hyperledger-sawtooth-intel-collaboration. Accessed: June 14, 2018.
[71] Burst, I. Q., 'Burst IQ', 2015. Retrieved from: https://www.burstiq.com/. Accessed: June 14, 2018.
[72] Azaria, A., Ekblaw, A., Vieira, T., and Lippman, A., 'MedRec: Using blockchain for medical data access and permission management', in *2016 2nd International Conference on Open and Big Data (OBD)*, Aug. 2016, pp. 25–30. doi: 10.1109/OBD.2016.11.

[73] Prisco, G., 'The blockchain for healthcare: Gem launches gem health network with Philips blockchain lab,' 2016. Retrieved from: https://bitcoinmagazine.com/articles/the-blockchain-for-heathcare-gem-launchesgem-health-network-with-philips-blockchain-lab-1461674938/.

[74] Kuo, T.-T., and Ohno-Machado, L., 'ModelChain: Decentralized privacy-preserving healthcare predictive modeling framework on private blockchain networks', *arXiv*, Feb. 05, 2018. doi: 10.48550/arXiv.1802.01746.

[75] Estonian eHealth, 'Health care blockchain,' 2016. Retrieved from: https://guardtime.com/blog/estonian-ehealthpartners-guardtime-blockchain-based-transparency. Accessed: June 14, 2018.

[76] Healthcombix, 'Healthcombix', 2016. Retrieved from: https://healthcombix.com/. Accessed: June 14, 2018.

[77] IBM, 'Blockchain in healthcare, patient benefits and more—Blockchain unleashed', *IBM Blockchain Blog*, 2018. Retrieved from: https://www.ibm.com/blogs/blockchain/2017/10/blockchain-in-healthcare-patient-benefitsand-more/. Accessed: June 14, 2018.

[78] ROBERT, Emilie, RIDDE, Valery, RAJAN, Dheepa, et al. Realist evaluation of the role of the universal health coverage partnership in strengthening policy dialogue for health planning and financing: a protocol. BMJ open, 2019, vol. 9, no 1, p. e022345.

[79] Hahnel, M., 'Blockchain enabled genome security from the moment it is sequenced', 2018. Retrieved from: https://www.genomes.io/wp-content/uploads/2018/03/The-genomes.io-Whitepaper-V-1.1.4.pdf. Accessed: June 14, 2018.

[80] Robinson, J., Robinson, J., and Leonard Kish, M., 'YouBase whitepaper,' 2016. Retrieved from: https://legacy.gitbook.com/book/joshrobinson/youbase/details. Accessed June 14, 2018.

[81] Kordestani, H., Barkaoui, K., Zahran, W., 'HapiChain: A blockchain-based framework for patient-centric telemedicine'. In *2020 IEEE 8th International Conference on Serious Games and Applications for Health (SeGAH)*. IEEE, 2020, pp. 1–6.

[82] Akhter Md Hasib, K. T. *et al.*, '[Retracted] Electronic Health Record Monitoring System and Data Security Using Blockchain Technology', *Security and Communication Networks*, vol. 2022, p. e2366632, Feb. 2022. doi: 10.1155/2022/2366632.

[83] Fu, J., Wang, N., and Cai, Y., 'Privacy-Preserving in Healthcare Blockchain Systems Based on Lightweight Message Sharing', *Sensors*, vol. 20, no. 7, Art. no. 7, Jan. 2020. doi: 10.3390/s20071898.

Index

G

H

I

J

K

M

N

P

Q

R

S

For Product Safety Concerns and Information please contact our EU representative GPSR@taylorandfrancis.com Taylor & Francis Verlag GmbH, Kaufingerstraße 24, 80331 München, Germany

Batch number: 10397790

Printed by Printforce, the Netherlands